THE HYDROGEN REVOLUTION

Praise for *The Hydrogen Revolution*

'Engaging, authoritative and very timely. Marco Alverà spells Hydrogen's critical role as an energy store in the clean power transition.'

Mike Berners-Lee

'To achieve the climate goals from the Paris Agreement, we need a wholesale transformation of our energy system. This book sets out compellingly the role that Hydrogen plays in this transformation and is an important contribution to advance the energy transition.'

Mark Carney

'An engaging and insightful overview of the tiny molecule that could revolutionise climate action. Like hydrogen itself, Marco Alverà is a superb connector – of ideas, approaches and practical, positive solutions.'

Dr Gabrielle Walker

'In this excellently-written and engaging book, Marco Alverà sets out an attractive vision for a hydrogen-fuelled future.'

Myles Allen
Director of Oxford Net Zero
Coordinating Lead Author of the IPCC
2018 Special Report on 1.5°C

'An invaluable explainer on hydrogen – a key to us achieving net zero . . . an urgent rallying call for action, a call policy-makers across the globe need to heed.'

Peter Mandelson

'This book presents a vision for the future based on hydrogen and renewables that is clear, grounded and hopeful.'

Francesco La Camera
Director General of IRENA

THE HYDROGEN REVOLUTION

A BLUEPRINT FOR THE FUTURE OF CLEAN ENERGY

MARCO ALVERÀ

BASIC BOOKS

New York

Basic Books
Hachette Book Group
1290 Avenue of the Americas, New York, NY 10104
www.basicbooks.com

Printed in the United States of America

First published in Great Britain in 2021 by Hodder Studio
First US Edition: November 2021

Published by Basic Books, an imprint of Perseus Books, LLC, a
subsidiary of Hachette Book Group, Inc. The Basic Books name
and logo is a trademark of the Hachette Book Group.

The Hachette Speakers Bureau provides a wide range of authors for speaking events.
To find out more, go to www.hachettespeakersbureau.com or call (866) 376-6591.

The publisher is not responsible for websites (or their
content) that are not owned by the publisher.

Typeset in Bembo MT by Hewer Text UK Ltd, Edinburgh

Library of Congress Control Number: 2021942288

ISBNs: 9781541620414 (hardcover); 9781541620421 (ebook)

LSC-C

Printing 1, 2021

Contents

Introduction   1

Part 1: A Hot Mess: Climate Change and How We Got Here   9
 1 A Net-Zero Goal   11
 2 The Day Everything Stopped   23
 3 Feet of Clay   27
 4 A Common Goal   37
 5 The Ups and Downs of Renewable Power   49

Part 2: Hydrogen: A 'How To' Guide   65
 6 Enter Hydrogen   67
 7 Flight of Fancy: Hydrogen's Early Uses   75
 8 The Allure of Oil   79
 9 The Hydrogen Rainbow: Extraction Methods   85
10 Handling Hydrogen   101
11 Using Hydrogen   113

Part 3: How Hydrogen Helps   119
12 Hydrogen and Electricity: A Power Couple   121
13 The New Oil   133
14 Making Materials Greener   145
15 Solving Seasonality   157

16 Green Lanes   167
17 Green Seas   183
18 Green Skies   189
19 It Is Rocket Science   201
20 Safety First   209

Part 4: Ready for Take-Off   215
21 The Mission   217
22 Catapult Companies   233
23 Calling the COPs   239
24 The Consumer Cavalry   255
25 Making Hydrogen Happen   259

Notes   263
Acknowledgements   275
Bibliography   277
Glossary   281
Appendix   289

To Lipsi and Greta who are made of starstuff

INTRODUCTION

We're in Venice, in the year 2050. It is the thirty-fifth anniversary of the Paris Agreement,[1] and it is a celebration. The spectre of climate change is finally exorcised. Global temperatures have started to stabilise, and we are set to feel the first refreshing drafts of global cooling. Rainforests and reefs survive – as does this beautiful city. Wildlife that was threatened is now returning.

We can trade, prosper and travel while respecting the equilibrium of our planet.

It isn't just our lighting and our electric vehicles that are totally green. Everything is. We use green fertilisers to grow our food. Our homes are connected to a green energy grid and produce, exchange and store their own clean energy. Ultralight vertical taxis bunny-hop over traffic jams. Long-haul aeroplanes leave contrails of ice in the sky. Boats, trucks, and buses glide noiselessly, no longer belching out CO_2 and fumes but emitting pure water vapour. The particles and pollutants that clogged up our cities and our airways are a thing of the past.

And all this because we are harnessing the power of the sun and the wind directly, and transforming it into hydrogen.

We are piping the desert sunlight and the ocean winds into our homes. Europe imports solar energy from North Africa and the Middle East, driving prosperity in the region. Australia harvests the strongest sunshine on Earth, then ships it to Japan on boats. China and India use Mongolian wind and Rajasthani sun to tackle the twin problems of air pollution and carbon emissions. The cost of energy is continuing to fall for everyone, helping to fuel economic growth and development around the world and create millions of jobs.

That's the dream.

Now let's return to reality.

Our efforts to solve climate change are way off track: 2020 was the second-warmest year on record, trailing only 2019. If we hadn't had Covid-19, 2020 would be remembered for the Australian bushfire disaster, which killed thirty-four people and more than three billion animals, and for the unprecedented wildfires in Brazil and California.

Climate change is also striking home for me, because I come from Venice. My hometown, ravaged by catastrophic flooding in 2019, has become a symbol of all that we have to lose from climate change and rising sea levels. Venice is not alone. Coastal areas around the world are threatened by rising sea levels. Cities are spending billions on flood defences. Extreme weather events cause untold damage to people, animals and livelihoods. Rising temperatures are threatening to make parts of the globe virtually uninhabitable.

I've long been concerned about climate change, but pessimistic about our chances of avoiding catastrophe. I was exposed to a healthy dose of climate science relatively early in my career as an energy executive, during a hike up a Norwegian mountain with climate strategist and author Gabrielle Walker. As we huffed and puffed (well, me mainly), Gabrielle explained how there was mounting scientific evidence that we were headed for disaster. It wasn't just a topic we should engage with intellectually, she argued. If there was even a chance the climate advocates were right, the risks were so great that we needed to do something about it now. It was like a Pascal's Wager* of energy.

She was right, of course, and it was the wake-up call I needed. I started to educate myself about climate change, and soon recognised that we were in very grave danger indeed. And then my daughters were born; since then, when I read about the predicted effects of unrestrained warming in 2100, I think about what it might mean for them and their generation.

The more engaged I became, the more my concern grew. I could see what fossil fuels were doing to the planet. What I couldn't see was how we could stop relying on them to power our industry, our travel and our trade. Yes, renewable electricity was making great strides, but electricity only accounts for 20% of our energy use. Even if we cleaned all of that up completely, using the sun and wind to generate clean electrons, we would still have the other 80% of the energy system to worry about. That's the energy we use in transport, industry and heating, which today rely mainly on molecules from coal, oil and natural gas.

We can switch some of those end-uses from molecules to electricity, of course. That's what we are trying to do with electric vehicles, and electric heating for the home. But there's

---

* Pascal posited that a rational person should live as though God exists, because if he or she does exist – even if there is only a small chance – there are an infinite amount of gains to be had and infinite losses to avoid by believing.

a limit to how much of the energy system we can switch to renewable electricity. Some sectors, such as heavy transport, industry and winter heating are particularly difficult for electricity to fully penetrate. The International Renewable Energy Agency (IRENA), sees electricity rising to just under 50% of the energy mix by 2050, which is wonderful, but still leaves another 50% to worry about. If we are serious about avoiding catastrophe, we need other technologies as well as renewable electricity – and we need them fast.

For many years, overcoming the enormous inertia in government, business and consumer behaviour felt impossible. But then, a few years ago, what looked set to be a boring business meeting changed everything.

I was in my office in Milan, in November 2018, almost at the end of a long day. As CEO of Snam, an energy infrastructure company with natural gas pipelines in Europe and the Middle East, part of my job is to think about what the global energy system of the future might look like, and what we might need to build to make it happen. Among my last appointments was the Snam scenarios team with a study that showed how Europe could reduce its CO_2 emissions to zero by 2050. The idea was to look at a host of clean energy sources – solar and wind power, biomass and hydrogen – and how much each might cost to produce, transport, store and use. Armed with this knowledge, a model could figure out the least costly combination of these sources in 2050.

As I looked through the study, I noticed there seemed to be a lot of hydrogen in 2050. An awful lot, really, for something that was almost absent from the energy mix and the policy discourse.

I had known about hydrogen's potential for a long time, ever since we'd made it from water in an experiment in science class at school, using one of those rectangular prism-shaped 9-volt batteries. I'd also encountered the dream of endless hydrogen energy when, at seventeen, I'd read Jules Verne's *The Mysterious Island*. In the novel, he talks about how 'water will one day be employed as fuel', and how the 'hydrogen and oxygen which constitute it, used singly or together, will furnish an inexhaustible source of heat and light, of an intensity of which coal is not capable'.[2] Reading Verne was how my slow-burn love affair with humanity's inexhaustible renewable molecule started.

Yet when I went to my first hydrogen conference in 2004, almost 130 years after *The Mysterious Island* was written, Verne's vision seemed as far away as ever. At the time, I was the head of strategy at the Italian utility Enel, and had been invited to Yokohama, in Japan, for a World Hydrogen Energy Conference. I had come back feeling that hydrogen still had a fatal shortcoming: it was wildly expensive. Whether you made it from nuclear power or from the nascent renewable electricity sector, it would cost much more than its fossil equivalent. I calculated that using hydrogen from renewables would bring the cost of a three-hour car journey to $4,000.

Yet, there we were, on that late afternoon in Milan in 2018, with a model forecasting that hydrogen was going to turn our energy system on its head. What did this new model know that we didn't? We ordered pizza and tried to work it out.

I'd learned a long time ago, as a young (and slightly overworked) Goldman Sachs analyst, that the answers models provide are only as good as their inputs. Garbage in, garbage out, as the saying goes. So, the inputs were where we started that evening in Milan, and we quickly got to the crux of the matter.

The model was forecasting that there was going to be a lot of cheap hydrogen for us to use because the cost of the renewable power used to make it was dropping fast, as was the cost of the equipment required to convert electricity into hydrogen. Transport costs were also assumed to be low because it would be delivered through already existing natural gas pipelines. All in all, the model predicted that by 2050 hydrogen would not only be the cheapest source of decarbonised energy for many sectors, it would actually be cheaper than what we are paying today for oil, coal and nuclear power.

That was a lightbulb moment, for me. I realised that hydrogen's true mission was to help us harvest sunlight and wind where they were in plentiful supply, transport them cheaply, and get them into our aeroplanes, factories and homes. Just 1% of the Sahara Desert gets enough sunlight to power the whole world[3] and hydrogen could finally give us a way to unlock that potential and decarbonise the hard-to-electrify sectors at the same time. Moreover, many people thought that the energy transition would mean rising energy costs and a need to support developing countries with billions of dollars. But the combination of cheap renewables and hydrogen meant we could envisage a net-zero world where energy was cheaper than it is today

The thought filled me with relief and excitement. If hydrogen really was a viable possibility – and we were still looking at a very big if – we finally had a clean molecule to deploy in the fight against climate change.

All we needed was a plan to deliver this vision.

That meeting fired the starting gun for our work on hydrogen. An intense few years of studies, field tests and projects were to follow.

When we started this work, there were few champions of hydrogen. But now momentum is building. There is growing consensus that hydrogen could account for up to a quarter of our energy needs in 2050.[4]

We are trying to get even more momentum going, not least through this book, a manifesto for this exciting new future of energy – a blueprint for how hydrogen can help save the world.

In it, I have tried to follow my own process of discovery, starting with the worrying science of climate change and the reasons why I was initially pessimistic about our chances of solving it. Thankfully, there are growing reasons for optimism – chief among them the spectacular rise of renewables, which can decarbonise swathes of the energy system directly and will provide the foundation of any solution.

Renewable electricity has limits, however, which means it cannot do the job alone. Hydrogen enables it to transcend these limits and can become the great energy connector, bringing together molecules and electrons, producers and consumers, countries and regions, and helping to get renewables into all the difficult corners of the energy world. We've long known about hydrogen's promise, but it is only now that we can see that it will become cost-competitive.

As well as explaining how hydrogen can help us get to net zero, I put forward a plan for how we can nudge it to its tipping point – the cost at which it will become competitive – faster, buying us precious time in the climate battle. This book lays out the steps we need to take – as businesses, policymakers and consumers – to unlock the power of hydrogen.

You can do a lot to help, through what you buy, how you vote, where you invest your savings, and the conversations you have. The thousands of seemingly inconsequential choices that we make every day can shape the world.

Part 1

*A Hot Mess: Climate Change
and How We Got Here*

I

A NET-ZERO GOAL

Climate change poses an imminent threat to our existence. We must reach net zero emissions faster than we ever thought possible. And we must do so while providing enough energy for a growing population and the development of emerging economies. This is no easy task.

In 1944, twenty-nine reindeer were corralled on a barge and floated north to what has to count as one of the most remote places in Alaska: St Matthew Island in the Bering Strait. They were intended as a roaming food source, in case the crew of a wartime radio navigation station on the island missed their supply shipments. When the Second World War ended in 1945, the men left and the station was abandoned. No one bothered to remove the reindeer though. They stayed put, breeding every year, until by 1963 there were fully 6,000 on the island. By 1964, the population had plummeted to just forty-three knock-kneed survivors. And even they didn't last long. There are no reindeer on St Matthew Island now. Every last one starved or froze to death.

In another part of the United States in the 1930s, three million tons of topsoil was blown off the Great Plains of America in a terrible dust storm. Daylight turned to darkness. Plagues of grasshoppers and jackrabbits descended to eat whatever meagre crops were left. Thousands of people died from inhaling the dust. Tens of thousands of poverty-stricken families, unable to grow crops, had to abandon their farms. Thee-and-a-half million people moved out of the Plains.

On Easter Sunday 1722, when Dutch explorer Jacob Roggeveen first set foot on Rapa Nui, he found an eerie sight. A barren, inhospitable land, with not a tree in sight. A remote rock in the middle of the South Pacific, buffeted by strong winds and salt spray, containing almost 1,000 massive and elaborate rock sculptures. Clearly, there had been a thriving community on the island at some point, with wood for logs to move big rocks, and time and energy to devote to the task. That was no longer the case.

What do these stories have in common? They're tales of ecocide, of environments so over-exploited that they destroy the very ecosystems that guarantee survival. The reindeer depleted their food source, lichen, faster than it could grow. The Dust Bowl was caused by over-ploughing and overgrazing the southern plains of the United States, leaving the topsoil defenceless against the winds. The story of Easter Island is contested, but at least one account sees deforestation as the key reason for its population crash.[1]

Could something like this await us, not as a community but as a species?

It is difficult to entertain the idea of one's individual or collective demise, but the risk was recently brought home to me in a casual conversation about Enrico Fermi, the

Italian–American physicist, and his famous paradox. 'But where is everyone?' Fermi is meant to have asked his physicist friends over lunch, referring to an earlier conversation about the high probability of intelligent alien life. If civilisations more advanced than ours exist, Fermi mused, why hasn't anyone been in touch? No one knows, of course. There's been lots of talk about technological disasters, nuclear Armageddons and the like. But what if some intelligent civilisations hit the limits of their environment, and faced an end akin to the reindeer on St Matthew Island? I hope that's not the case. Nonetheless, that thought did give me a chill, and highlighted the sense of urgency with which we must approach global warming.

We are certainly pushing up against the limits of what the planet can countenance. Waste and pollution are poisoning the very air we breathe. According to the World Health Organization (WHO), ambient air pollution kills more than 4 million people every year – twice as many as Covid-19 killed in 2020, and twice as many as are killed every year by malaria and tuberculosis combined.[2] Today, millions of species are threatened and many have already perished. Scientists are calling it the sixth mass extinction because it follows five others that all took place between 440 million and 65 million years ago. But this time, the mass extinction underway isn't due to a volcanic eruption or to a collision with an asteroid. This time it's down to us.

Burning issue

Life on Earth depends on a delicate balance of gases. Plants absorb CO_2, using the carbon to make trunks, shoots and leaves, and then release the oxygen. We, like all animals, eat the carbon (pasta is a mixture of carbon, oxygen and hydrogen). We inhale oxygen to break up the food and release energy, and exhale CO_2. When this cycle is in balance, the amount of CO_2 in the atmosphere is stable. But it is now out of whack.

If you cut down a forest, the carbon that was stored in the trees all gets into the atmosphere (whether the wood burns or rots). Burning fossil fuels releases carbon that was trapped by plants and animals that lived millions of years ago. Once released, CO_2 will stay in the atmosphere for centuries.

Fossil fuels are by far the biggest source of CO_2 emissions, accounting for 33 billion tonnes (gigatonnes) a year in 2019.[3] Other CO_2 emissions come from industrial processes and changes in land use (like cutting down forests) bringing the total for extra CO_2 to be injected in our atmosphere to around 40 billion tonnes.[4] On top of that, you will also hear about CO_2 equivalent emissions, which include other greenhouse gases such as methane (produced for example by rotting plants and cows' stomachs, or as the result of fugitive emissions from natural gas production). Methane has an even stronger impact on climate change than CO_2, albeit on a shorter timescale. If you convert the other greenhouse gases to CO_2 equivalent, and add it all up, you get to total emissions in the region of 52 billion tonnes of CO_2 equivalent.[5] This book will focus on CO_2, and won't look at land use, but bear in mind that to get to zero we'll need to cut emissions of other greenhouse gases too, no easy feat as it will require lifestyle changes.

Where do CO_2 emissions come from? The following figure shows the breakdown by sector.

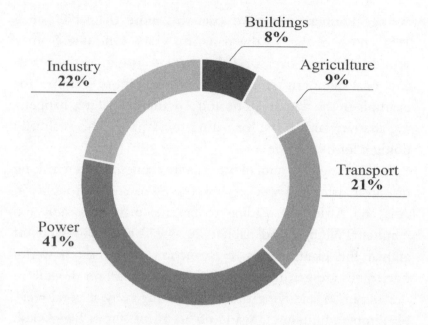

Total estimated CO_2 emissions by sector, 2019[6]

Because of our actions, the concentration of CO_2 in our atmosphere has now built up to about 415 parts per million, almost double the amount before the Industrial Revolution. That is far higher than at any time in the last 800,000 years at least, and it is still rising rapidly.

Having some CO_2 and other greenhouse gases in the atmosphere is a necessary thing. They absorb heat coming up from the Earth's surface and emit some of it back down again. We need this trapped heat to survive. Without it our planet would freeze.

But now we have too much. Excess carbon in our atmosphere is trapping more and more heat. It doesn't matter who emits it, or where in the world it gets hurled into the air: once it gets there, it is everyone's problem. As a result, global average temperatures have risen by more than 1°C since 1880, most of that in the past fifty years. One degree may sound small, but over land the increase is higher than that, and local swings in temperature are much more extreme, for example in the fragile Arctic and Antarctic. And it is happening at a rate far too fast for nature to adapt to. This is already doing a lot of damage.

Across the globe, rainfall patterns are changing, often making dry areas drier and wet areas wetter. Worsening droughts in Asia and Africa are leading to famine, mass migration and conflict. Pollution and habitat loss are driving thousands of animal and plant species to extinction. The most powerful hurricanes are getting stronger. The extra carbon dioxide in our oceans is acidifying them, and eating away at coral reefs, plankton and molluscs. Sea levels are rising. From Bangladesh to Manhattan, hundreds of millions of people living close to sea level could be displaced.

Sophisticated climate models, based on the known physics of the atmosphere and oceans, predict planetary warming by about 4°C by 2100 if emissions are unrestrained. This would make large areas of the planet inhospitable to humans – but that's not the worst of it.

Disasters have a habit of unfolding gradually, then all of a sudden. And global warming is no different. Once temperatures rise too far, we will hit tipping points, where things suddenly get a lot worse. These tipping points include the Amazon rainforest drying out and dying off, and the Greenland and West Antarctic ice sheets collapsing, bringing a total sea level rise of more than 10 metres (albeit over a few centuries). Worst of all, marine sediments and thawing permafrost could release huge amounts of the potent greenhouse gas methane, heating the planet further. This could be a truly fatal tipping point, rendering the Earth almost uninhabitable.

We may already have passed some of these tipping points – the relatively mild matter of collapsing ice sheets. Others could be reached soon, if the world keeps warming

The hero is zero

To avoid catastrophic climate change, the consensus is that we should try to stay below warming of 2°C, and ideally below 1.5°C. That doesn't leave us much room for manoeuvre. Even if emissions stop rapidly, temperatures will keep rising for a while. That's because the oceans have a tremendous capacity to hold heat; they take a long time to warm up, which in turn keeps the air slightly cooler for a while.*

It is not good enough to merely reduce emissions. That's because when we release carbon dioxide into the atmosphere, it lingers for a long time. Even after a century, a third of that gas is still up there; after a thousand years, nearly a fifth. This means the carbon build-up that's causing global warming is a little like a kitchen sink (with the plug in) that is about to overflow. Even if you slow down the flow from the taps, the water level in the sink is still rising. And even if

---

* Adding greenhouse gases is just like turning up the heating; its effect is measured in watts per square metre of the Earth's surface. And when you switch on a heater in a cold room, the air might warm up by 10 degrees in a few minutes; but the walls have a lot more capacity for heat and so they take longer to warm up. While the walls are still cold, they are cooling the air a little; as they slowly warm, the air temperature will keep creeping up until it hits a steady final temperature, when everything reaches equilibrium.

you stop the flow, it will take a very long time for the water to evaporate.

Climate scientists have worked out roughly how much more CO_2 we can emit without driving temperatures above 1.5°C. To give us a 50–50 chance of staying under that threshold, the answer is around 440 gigatonnes.[7] To stay below 2°C, the budget is around 700 gigatonnes. This is our budget for CO_2 alone, and assumes a rapid decline in other greenhouse gas emissions. There is also quite a lot of uncertainty about exactly how the world will respond to the increased quantities of CO_2 in the atmosphere, and some estimates for the carbon budget are lower than this.

If you take the budget and divide it by our CO_2 emissions of around 40 gigatonnes, that leaves us about eleven years at current rates before we break the 1.5°C mark, and something like eighteen years before we hit 2°C. All of this is from my perspective writing in 2021.

And that's at current rates. Bear in mind that emissions may still be increasing. In 2019, energy-related emissions rose by around 1%. In 2020, we changed course in the face of Covid-19. With people working from home, and transport slowing or stopping completely, energy-related emissions dropped by 2 gigatonnes.[8] Much of this fall is reversing as life returns to normal. Just for reference, at the time of writing, the International Energy Agency was forecasting a rebound of 1.5 gigatonnes of CO_2 emissions in 2021,[9] bringing us almost back to the 2019 peak.

To stay within budget, we need our net carbon emissions to fall to zero. How soon we have to do this will depend on the path we take. Assuming a simplistic linear fall in emissions, we would need to hit global net zero by the early 2040s to stay within 440 gigatonnes. If we can cut emissions at a faster rate sooner, that would buy us more time. In carbon budget terms, a tonne of CO_2 cut from annual emissions today is worth ten times a ton cut in ten years' time. It would help if we could also aggressively cut non-CO_2 sources of warming such as methane emissions.[10]

We can also actively remove carbon from the atmosphere, by planting trees, burying charcoal, grinding up rocks or using solvents to capture CO_2 from the air, a process known as direct air capture (DAC). If, as well as cutting emissions, we are also doing some of these things, that will reduce net emissions faster and stretch out the carbon budget. Indeed, three of the 1.5°C-compliant scenarios that the Intergovernmental Panel on Climate Change (IPCC) has published use carbon removal to square the numbers on a very large scale.[11] But carbon-negative technologies are not an excuse to keep emitting with abandon. They are either limited in scale or unproven – or both. So, the imperative is to cut emissions – as much as possible, as soon as possible.

This implies a revolution in the way we produce and use energy.[12] Today, we use fossil fuels for 80% of our energy. The remaining 20% is clean power, but you should bear in mind that most of that is biomass, nuclear and hydropower, which have been around for a long time and whose growth prospects are, for many reasons, lacklustre. The new renewables of wind and solar currently make up 2% of our primary energy (a global average; some places have a significantly higher penetration of renewables).

Getting to zero will mean using mainly solar and wind power, rather than coal, oil or natural gas, for power generation; changing our vehicles so they no longer use oil; modifying materials and the industrial processes we use to make them so that they don't require fossil fuels as feedstocks; and heating our homes without fossil fuels.

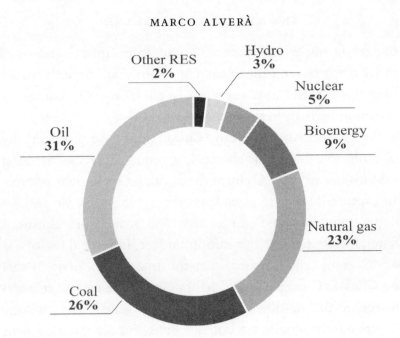

Global primary energy demand by fuel, 2019

What's more, we need to transform our economies at a time when both demographics and development will put additional pressure on the energy system. By 2050, there will be around 9.7 billion of us globally, up from 7.9 billion today. How much energy each of us uses also tends to increase with prosperity. For an idea of the disparity that exists in the world today, take a look at the chart below.

Table 1: Energy consumption per capita, 2019

Energy consumption across the globe	MWh/year
Global average	22
USA	78
EU27	35
Africa	8
Asia Pacific	17

In 2019, US citizens consumed nearly 80 megawatt hours (MWh) of energy per person – compare that with around 8 MWh in Africa. Globally we consume some 170,000 terawatt hours (TWh) of energy per year today. That will be even greater by 2050, and it will all need to be clean.

Given the scale of the challenge, it's no wonder that some people reckon we should just . . . stop.

2

THE DAY EVERYTHING STOPPED

The global pandemic and the ensuing global lockdowns showed that we cannot solve the climate crisis through lifestyle changes. There is room for individual action, but we need a solution that combines net zero emissions and life as we know it.

The call to stop everything has its value – as a gesture. It concentrates minds. It makes people pause and think, which is, let's face it, never a bad strategy. People criticise Extinction Rebellion, the global movement advocating civil disobedience to urge climate action, for its theatrical gestures. *Stop burning stuff! Stop flying! Stop making and buying things! Stop eating meat! Stop having kids!* People who complain about such theatrics are forgetting the power of theatre. You don't have to take something literally to take it seriously. There is such a thing as a climate emergency, and we do need to re-examine many of the things we've taken for granted.

The decision was taken out of our hands in early 2020, when Covid-19 became a global pandemic. And everything did stop. We stopped going to workplaces. We stopped driving. We stopped flying. We stopped going to cafés and restaurants. We stopped sports. We stopped schools.

And it was miserable, especially for the poorest and most fragile and marginalised people in society, who have suffered from job losses, company bankruptcies and rising inequality.

So, no. Stopping all industry, travel and commerce isn't going to work. Not because vested interests will suffer, but because *people* will suffer. And, even if we could force such an iniquitous solution on the world, it wouldn't do the job.

The lockdowns that caused such misery did give us a glimpse of the quiet, clean world we could be enjoying: fresh air in our cities; cyclists and pedestrians enjoying roads free of traffic. I marvelled at those pictures of wildlife taking over the streets of capital cities globally, and saw first-hand the fish shoaling in the clear waters of Venice's canals (although that may have less to do with pollution and more to do with the fact that there were no boats to stir up the sediment). And the lockdowns did have some impact on pollution and CO_2 emissions. According to satellite observations by Europe's Copernicus Atmosphere Monitoring Service, China saw a 30% drop in two key air pollutants, nitrogen dioxide (NO_2) and particulate matter, over the month of February. In Italy, in March 2020, those same pollutants fell by 40 to 50%.

But it was nowhere near enough. In 2020, CO_2 energy-related emissions came in 6% lower than 2019.[1] Say that we continued to cruelly stunt our economies, Covid-style. We could expect our emissions to stay at 2020 levels – still far short of net zero. Even a painful shambling version of business as usual will tip us into irreversible climate change.

So clearly 'just stop' is not going to work.

That doesn't mean we can't learn from the pandemic. Disruptions provide opportunities to reflect, rethink and change things. This seems to have happened, at least to some extent. I have been encouraged by the rapid and relatively cohesive response by policymakers in Europe, who, faced with the prospect of having to pour money into the economy to prop up the system, wisely decided to try to combine stimulus measures with green objectives. Of the €750 billion of funds earmarked as recovery and resilience, 37% has to be tied to climate change projects. In the US, too, President Biden unveiled a $2 trillion infrastructure bill, which he called a 'once-in-a-generation investment in America', including spending on roads, bridges, ports and railways, and also measures to encourage the uptake of electric vehicles and renewable power.

The 2020 pandemic has provided painfully clear evidence that we can't 'just stop', but it has also handed us a chance to rethink our lives and build back better. Will we be able to seize this opportunity? So far, our track record on really, seriously tackling climate change has been woeful – but things do seem to be turning around.

3

FEET OF CLAY

We have long known about climate change, but we haven't done well at tackling its causes. We've lacked a clear goal, and available technologies have been limited and expensive, setting climate change mitigation at odds with economic development.

'. . . at the rate we are currently adding carbon dioxide to our atmosphere, within the next few decades the heat balance of that atmosphere could be altered enough to produce marked changes in the climate.'[1]

Guess when that was written? 1966. By Glenn T. Seaborg, Nobel Prize-winning chemist and chair of the US Atomic Energy Commission. Scientists have been warning of the danger for many decades.

It's been over twenty years since the Kyoto Protocol was adopted,[2] with 192 signatories (the US signed it but did not ratify it), committing industrialised economies to reducing greenhouse gas emissions. Yet since then, almost 700 gigatonnes of CO_2 have been hurled up into the atmosphere. That's not far from the amount emitted since the beginning of the Industrial Revolution up to that time. At current rates, over the next thirty years we'd emit as much as we did for the past 250 years.[3]

As Jonathan Franzen put it: 'The struggle to rein in global carbon emissions and keep the planet from melting down has the feel of Kafka's fiction. The goal has been clear for thirty years, and despite earnest efforts we've made essentially no progress towards reaching it.'[4]

Why have we – humankind – done quite so badly?

Well, it hasn't helped that we lacked a clear and positive goal. For a very long time, the whole climate change narrative was simply about reducing emissions by this or that percentage, a negative goal. Designer William McDonough has described this as being a little like getting into a taxi and telling the driver 'I'm not going to the airport' or making a resolution to 'reduce one's badness'.[5]

The main reason we lacked a positive goal was that we didn't quite know what to do. We had no good, or even reasonable, solution to hand. The key problem was that our main tool to fight climate change was renewable electricity, and at that time renewable electricity was much, much more expensive than fossil fuels.

That meant that even small cuts in emissions came at high cost. To ram it into the energy system, the early adopting countries had to pay billions in subsidies. Astonishingly, the first solar auctions in Italy paid over €450/MWh for solar power for twenty years, on top of power prices (for reference at the time they were in the region €60–70/MWh). If you assume that this renewable electricity displaced power produced by natural gas, every MWh produced by renewables avoided 370 kg of CO_2 emissions. Divide the premium paid by the emissions avoided, and you get €1,200/tonne of CO_2. For an idea of just how big a number that is, just think that today Europe puts a price on carbon emissions in some sectors (the

Emission Trading System), and that price is not far off €50/tonne.

As a result of this early and wildly expensive push into renewable power, European countries committed to something in the region of €750 billion in subsidies. In Italy we are still paying €11–12 billion a year, which adds €75, or 15%, to the average household's annual electricity bill. For governments, policies that impose high energy costs are problematic. Energy poverty is a growing phenomenon, and adding extra charges on to utility bills is an unfair way to raise funds because it affects the less well off – who spend a higher proportion of their disposable income on energy – more than the wealthy. Curbing emissions by raising the price of energy may make excellent strategic sense, but a nation is made of people and a 10% tax on the price of petrol for some may mean having to reduce basic needs elsewhere. In France, in the last couple of years, the protest movement known as the *gilets jaunes* (yellow vests) has taught politicians a valuable lesson in how hard it is to raise energy costs even by small amounts.

This is the conundrum that was facing governments: they somehow needed to transfigure industry and the economy in such a way that people didn't lose their jobs as economies got hit by high energy prices. And until very recently, there didn't seem to be a strategy to avoid climate change that didn't also involve impossibly painful trade-offs.

The main reason why we lacked a positive goal was that we didn't have the underlying technology to hand. But it didn't help that the energy industry is so complex and fragmented, and struggles to cooperate across sectors.

The problem here is that different operators have limited

awareness of what everyone else is doing. For instance, the electricity sector knows a lot about gas used to generate power, but less about gas used in heating, industry and transport. Gas companies see the power sector as a client, but do not spend much time analysing the challenges of balancing a grid. For many years, that didn't matter so much. Companies identified the needs of the sectors they operated in, and solved them as best they could. All worked swimmingly well, for as long as coal, oil, gas and electricity were produced and consumed separately.

Today, there are efforts to cooperate across sectoral lines. But these efforts are hampered by the fact that there is no one coherent way of measuring and describing energy across sectors. Other industries have consistent units. Information and communications technology for example, just uses bytes (kilobytes, megabytes, gigabytes, terabytes) for stocks of data and bits per second for flows; car companies use horsepower. Having one measurement system makes it easier to choose a computer, an internet connection or a car. At least you're generally comparing like with like. If you need energy, however, prepare to drown in alphabet soup: you can measure it in joules, or gigajoules (GJ); electricity companies think in megawatt hours (MWh); oil producers deal in barrels of oil equivalent (boe) or tonnes of oil equivalent (toe); gas companies see the world in standard cubic metres (scm), or cubic feet, or millions of British thermal units (MMBtu). Mining companies measure tonnes of coal equivalent (TCE). Climate scientists chart gigatonnes of CO_2-equivalent emissions ($GtCO_2e$). And do you want to measure capacity, or hourly, daily or yearly flows?

Thinking about a full-system pathway for climate change, and which technologies might be able to do what, is a bit like trying to choose a T-shirt on the internet when you can see the pictures but can't quite work out what size each shirt is, how many per pack and what they cost.

If you want to be able to compare different fuels to put them in the same 'currency', what you need is a version of the pocket conversion chart which I carry around and consult every time someone mentions a unit that's outside my comfort zone. The table below shows you the relationship between the different units (so 1 MWh – which is very roughly the electricity used by an Italian family every four months – is the same as 0.6 of a barrel of oil, or 91.4 standard cubic meters of gas, and so on), and also the dollar amount that you'd be paying for 1 MWh of energy in different fuels. In the first three months of 2021, 1 MWh in oil cost $38, while 1 MWh in gas cost $22 in Europe and $10 in the US, meaning oil was 1.7 times more expensive than natural gas in Europe on a per energy basis, and almost four times as expensive as natural gas in the US. For more detail on hydrogen energy calculations, please refer to the appendix on page 289.

Table 2: Energy unit conversion

Energy (MWh)	Oil (boe)	Natural gas (scm)	Natural gas (MMBtu)	Coal (TCE)	Hydrogen (kg)
1	0.6	91.4	3.41	0.12	25

Table 3: Energy costs per MWh in the first quarter of 2021

	Oil Global	Natural gas Europe	Natural gas USA	Coal Europe	Grey hydrogen	Green hydrogen	Blue hydrogen (Europe)[6]
Energy equivalent costs ($/MWh)	38	22	10	14	50	100–140	60

Complexity and confusion make it difficult to identify a pathway. And this is all the more worrying because the usual pace of change in the energy system is glacial.

Most people think that the nineteenth century was dominated by coal and the twentieth century by oil and that the twenty-first will belong to renewables. But the nineteenth century didn't run on coal. It ran on wood, charcoal, and cereal straw, which accounted for 85% of the world's energy. And for most of the twentieth century the biggest energy source wasn't oil, it was coal. Crude oil didn't surpass coal until 1964. And even today, cheap natural gas in the US hasn't supplanted diesel in trucking, despite the fact that it costs less.

This inertia is not a modern malaise, a political wrong-turn that we might easily put right. Rather, it's a reflection of just how difficult it is to change things.

Add together the limited toolkit and the fragmented industry, and it isn't hard to see why, for so long, climate change efforts lacked a North Star; an aspiration for everyone to aim for.

This translated into a lack of momentum at the grassroots level. Think about it from the perspective of people going about their daily lives. On the one hand, they had an inkling that we might, at some point, be facing catastrophe. On the other, they couldn't see a solution that didn't call for unimaginable sacrifices.

What do people do in this sort of context? In general, they look away. As George Marshall pointed out in *Don't Even Think About It*, his book about how our brains are wired to ignore climate change, '. . . it's hard to look too long at the thing that causes anxiety'.[7] We all strive for normality. As

Jonathan Franzen puts it, people prefer to think about break-fast than about death.[8] In that sense, climate change is a particularly prickly problem: we feel disempowered; it's so huge, it seems too difficult to overcome.

Without momentum, cohesiveness is hard to achieve. Take the analogy of a bicycle. At speed, it moves in an elegant, forward motion (and is quite easy to ride). If it is barely moving forward, it wobbles all over the place. Crusades and campaigns are a bit like that. When they don't have momentum, the parties that are meant to be working together grow distrustful and bicker. Climate change becomes politicised – which has been especially true in the US. Environmentalists and the energy industry can find no common ground.

Countries, too, have in the past failed spectacularly to come together in the face of a common threat, playing a game of beggar-thy-neighbour in hopes that the cost of decarbonisation would fall elsewhere. For years the United States refused to sign the Kyoto Protocol – a heavy blow given it accounted for around a quarter of global emissions by itself. President George W. Bush resisted on the grounds that it would damage the US economy and was unfair because it did not include developing nations such as China and India.[9] People in the developing world didn't see it quite like that, given they had done little to cause the problem in the first place; they are only now increasing their energy consumption as living standards climb. They didn't take kindly to the idea of a global agreement that would increase their energy costs, erode their competitive advantage and limit their growth potential.

I saw the lack of cooperation first-hand when I went to COP24, the twenty-fourth meeting of the Conference of the

Parties of the United Nations Framework on Climate Change, which was held in Katowice in south-west Poland, the coal capital of Europe, in 2018. I have rarely witnessed a more depressing sight. Hundreds of negotiators, working through the night for weeks on end and getting nowhere. Over lunch, a charming representative shared his views on the debacle. Apparently, the top diplomats of an influential Gulf country were flat-out refusing to play ball, while the Americans – who had hitherto been the main purveyors of carrots and sticks to bring recalcitrant countries into line – were barely present. And without the Americans on board, many were now dragging their heels. There was no possibility of solving any element of this deadlock at the actual talks, so it was just an exhausting grind with no light at the end of the tunnel.

I think the litmus test of just how difficult it is for the world to cooperate on climate change is the issue of a global carbon price. As a policy, it would have two key advantages. It would let the market determine where emissions could be abated most cost-effectively – globally, which would be a lot more efficient than every country setting its own targets. And it would raise money that could finance the transition and be funnelled to cushion its blow for those who would suffer most from it – such as workers in coal-producing regions. As a former economics student, I find this idea attractive. But I suspect it isn't going to happen. Just saying the words 'global carbon price' is enough to stop any meaningful debate – senior policymakers tend to switch off, muttering about impossible solutions.

That said, you don't necessarily need a global carbon price to get significant policies going. Coalitions of the willing, bilateral agreements with a whiff of horse-trading and even unilateral action can all push the world closer to decarbonisation. The EU, for instance, has an internal carbon price, and is considering a carbon border tax (in effect charging countries who want to import goods according to the amount of carbon that was emitted in their manufacture). Over time, such an approach may convince other countries that it might be better for them to impose their own internal carbon price and skip the EU border adjustment. This is one of the ways in which we might be able to get a system of loosely coordinated carbon prices around the world.*

Since Paris, successive COPs have failed to be the springboard for action that we need. In a context in which we've had no clear goal, an insufficient technological toolkit, and a lack of leadership and momentum, that's not terribly surprising.

But we really are up against it, with 1°C of global warming under our belt already, and temperatures heading speedily upwards. We need 2021 to go down in history as the turning point for climate, the time when we really grasped the nettle, making good use of the trillions of dollars pouring into the economy post Covid.

Thankfully, much of the momentum we've been missing seems to have returned.

---

* China introduced a carbon market in 2021 – which marks a significant step on the road to a global carbon price.

4

A COMMON GOAL

Don't despair. We now have a goal: net zero is providing clarity, a sense of purpose, and catalysing action. A new generation is committed to the climate cause, and the political landscape has shifted. Green investments are pouring into renewable energy, which has become remarkably cheap. And it all happened really fast.

I am feeling more optimistic about our chances of solving climate change than ever before. After decades of global dithering, things seem to be moving in the right direction. I really do think we are at an inflection point – there are so many reasons to be hopeful, and so many signs of change.

For one, we now have a clear goal. Net zero is all the rage. At the time of writing, we have net zero by 2050 commitments from the EU, the UK, Norway, New Zealand, Japan, South Korea, Chile, South Africa, Switzerland and Costa Rica. Even more startlingly, China has pledged to drive down emissions to virtually zero by 2060. The importance of this statement cannot be overestimated, as China alone accounts for 29% of global emissions. America, too, seems well on the way to giving itself a net zero target, and has already committed to reducing emissions by 50% below 2005 levels by 2030.

Net-zero thinking has done us a huge favour: it is focusing minds in a way that vague global temperature goals have failed to do in the past. In Europe, we have moved from having to allocate EU-wide targets to the different countries and then across sectors, to a simple and bold target that even my young daughters are fond of.

Net zero doesn't change what we have to do, but it leaves us with no place to hide. If we have to eliminate emissions by 2050, then everyone has to do their bit. No company can argue that their business will be unaffected. Voters, investors and consumers will increasingly be able to hold states and companies to account. It also changes the narrative – from negative, limited, costly and unfair to something that sees us all pitching in together, and actually has a sporting chance of success.

Having a target nudges us closer to a plan. To get to zero, we will need to design our solution with the end in mind, moving away from an 'every little helps' mindset where we just tackle the next easiest problems, to something more long-lasting.

As well as having a broadly accepted goal that everyone can get behind, we now have charismatic 'climate leaders' in all walks of life.

The arrival of Joe Biden in the White House, to replace Donald Trump, is a game-changer. One of his first acts was to re-join the Paris Agreement, which President Trump had ditched. He has also signalled that the US aims to play a leading role in upcoming climate negotiations by hiring John Kerry as climate envoy. The 2015/2016 Paris Agreement benefited from the political heft of Kerry, who at that time was President Obama's Secretary of State. As President Biden's climate envoy, he now gets to crack the whip in a job he knows well. As he seeks to extract commitments, countries may be more responsive than they have been in the past, especially those that curried favour with the previous US administration and now have to scramble to prove their credentials with the new one.

America's renewed commitment has helped reinvigorate China's climate leadership, too. President Xi Jinping's collaboration had been instrumental to getting the Paris Agreement signed, but then China's enthusiasm seemed to wane. Now things have changed again. At a climate summit in April 2021, he vowed to 'phase down' the use of coal from 2026, and has promised to collaborate with the US on the climate agenda despite the strained relationship between the two superpowers.

Elsewhere, the EU seems keen to preserve the competitive advantage it has built on climate. The goal of reducing emissions by 55% compared with 1990 was enshrined in law in April 2021. And the UK is aiming to cut 2035 emissions by 78% compared with 1990.

That politicians are showing more mettle is at least in part down to a seismic shift in how society views climate change. It helps that the kids are – finally – in charge, or at least old enough to protest and march while cajoling and shaming their parents into action. The people who will be around for the brunt of climate change, those who really are scared for their lives, are now able to influence the political agenda. Greta Thunberg has gained real traction as the spokesperson for climate change, helping to catalyse grass-roots momentum, starting with those still in school.

I am exposed to the budding environmental conscience of my two girls every day. The pre-Covid school run was a chorus of 'Why don't you get an electric car, daddy?' (I have since bought a hybrid), 'Why do we have plastic lunchboxes?' (now replaced), 'Why are there so few trees in Milan?' (Snam has founded a company that plants trees to offset CO_2). One reason I developed an urgent desire to help tackle climate change was my concern for my daughters.

Entrepreneurs have also done a lot to capture people's attention and make the renewable revolution cool. That electric cars are no longer seen as boring and utilitarian is largely down to Elon Musk, whose hi-tech Teslas are the stuff of consumer dreams (as well as being the prerogative of the very wealthy).

Changing attitudes still need to translate into coordinated action. We will need to overhaul how we make things, how we move things and people around, how we keep warm and cool. *The Financial Times* has said that decarbonisation will require a wartime level of mobilisation[1] – and it is undoubtedly right.

So how will we get the 8 billion people who live on the planet to coordinate and pull in the same direction?

As I learnt from reading Yuval Harari's book *Sapiens*,[2] what sets man apart from other animals is the capacity to cooperate with people we've never seen, through shared beliefs and mental constructs which shape how we collectively behave. Money and markets are among the ways in which we do this.

The rush for green investments is a case in point. Today, investors and savers want their fund managers to pick green stocks. They think it is good business, because companies on the wrong side of history will lose market share or even go out of business, while those that are leading the change will have lots of opportunities. They also think that by rewarding the goodies with cheaper capital, and starving the baddies of money to invest, they can help shape decisions that companies will make.

As a result, big funds are becoming more selective about what they invest in. They are also becoming more vocal. By 2020, more than 3,000 organisations representing more than $100 trillion in assets had signed up to the Principles for Responsible Investments – a UN-backed initiative holding that environmental, social and governance (ESG) factors must be taken into account when making investment decisions. Some go further. Aviva Investors, for example, which manages £355 billion, has warned that it will divest from thirty oil and gas companies unless they do more to tackle climate change.

The 800-pound gorilla in the asset management world is BlackRock, which manages $8.7 trillion in assets and is a shareholder in the world's largest companies. Larry Fink, its chief executive, wrote in 2021 that 'no issue ranks higher than climate change on our clients' lists of priorities. They ask us about it nearly every day'. That comment appeared in a letter he wrote to the chief executives of companies BlackRock had invested in. 'We are asking companies to disclose a plan for how their business model will be compatible with a net zero economy.' Fink warned that a lack of progress would lead to BlackRock voting against management at shareholder meetings, and potentially ditching the stocks.[3]

Companies need a way to tell the market, clearly and transparently, just how green their business is, which means having a shared view of what counts as green in the first place. One scheme that will help is the EU taxonomy: a classification of different business activities and their level of environmental compatibility.

Meanwhile, companies with evident green credentials are doing very well. They are growing fast, investing in ever bigger green projects and being rewarded by the stock market. The market capitalisation of NextEra, a US utility focused on solar and wind power, overtook that of the world's largest oil and gas company, Exxon in 2020. Another clean energy darling is Danish company Ørsted, which used to be an oil and gas business called DONG Energy. Since 2016, it has sold all of its fossil business and opted to create a world-leading renewable business instead. In 2021 its stock was worth over four times what it had been five years earlier. These clean utilities are now called supermajors, a name that until recently was reserved for big oil companies. Tesla, too, was worth more than $650 billion in April 2021.

Unsurprisingly, companies on the wrong end of the trend are rushing to adapt. Oil and gas majors did not have a good year in 2020. The five biggest companies saw a combined $77 billion wiped off their balance sheets. Admittedly, it wasn't a normal year, as Covid-frozen economies didn't use much oil. But compared with the stellar performance of some renewable energy stocks, it hammered the point home. It is no accident that companies such as BP, Shell and Total are trying to reduce their carbon footprints and increase the share of their business that comes from low-carbon and renewable sources.

Big companies outside the energy sphere, too, are getting behind the cause. Since 2019, dozens of multinationals have committed to net zero. Apple by 2030, Unilever by 2039, Amazon by 2040. Microsoft has gone one better, committing to be 'carbon negative' by 2030. Sony has threatened to pull out of Japan because of the country's lack of renewable infrastructure.

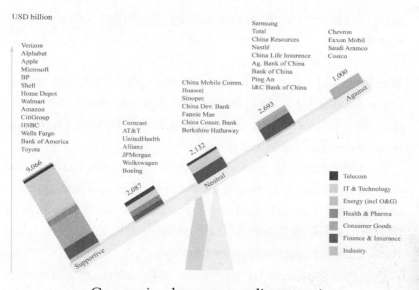

USD billion

Verizon
Alphabet
Apple
Microsoft
BP
Shell
Home Depot
Walmart
Amazon
CitiGroup
HSBC
Wells Fargo
Bank of America
Toyota

9,066

Comcast
AT&T
UnitedHealth
Allianz
JPMorgan
Wolkswagen
Boeing

2,087

China Mobile Comm.
Huawei
Sinopec
China Dev. Bank
Fannie Mae
China Constr. Bank
Berkshire Hathaway

2,132

Samsung
Total
China Resources
Nestlé
China Life Insurence
Ag. Bank of China
Bank of China
Ping An
I&C Bank of China

2,693

Chevron
Exxon Mobil
Saudi Aramco
Costco

1,000

Supportive   *Neutral*   *Against*

- Telecom
- IT & Technology
- Energy (incl O&G)
- Health & Pharma
- Consumer Goods
- Finance & Insurance
- Industry

Companies that support climate action
Image © the Rocky Mountain Institute

I am encouraged by this chart, which shows that companies that explicitly support climate action have a higher aggregate market value than the more passive ones.

The hope is that the political and corporate worlds will egg each other on. As governments negotiate domestic and global climate targets, they will increasingly benefit from a business community that has shifted from being somewhat sceptical, and very concerned about the costs of the energy transition, to one in which many companies have already run a couple of laps in their race to net zero.

Many of the pieces of the puzzle are now in place. We have a goal, climate leadership, societal momentum and green funding. The last piece of the puzzle, and the foundation for everything else, is the recent surge in renewable power. Previously expensive and niche, it is now cheap and fast-growing. That's why we can now truly get excited by the prospect of a net zero world.

Sunshine and wind are tantalising resources. They are generously supplied for free by nature, but you can't use them directly – you need specialised technology to convert light and movement into electrical power.

In 1839, French physicist Edmond Becquerel discovered that some materials generate an electric current when exposed to light, but it wasn't until 1940 that we saw the beginnings of modern solar cell technology, when Bell Labs researcher Russell Ohl coaxed a current out of silicon.

For a long time, solar power was eye-wateringly expensive. In 1956, the price for solar panels was more than $200 per watt of capacity.[4] That's a lot – the solar power that you'd get from it would cost something like $20,000/MWh. As a result, solar power was out of reach except for some niche applications. One wasn't even on this planet. In 1958, the US launched *Vanguard 1* – just the fourth artificial satellite, and the first to be solar powered. Space agencies could afford to pay up, but these prices were no good for everyday applications.

Solar eventually came back down to Earth thanks to the work of countless scientists and engineers, such as US photochemist Elliot Berman. His insight was that photovoltaic (PV) cells could be made using castoffs from the semiconductor industry. Imperfect silicon was of little use in electronics but worked fine in solar cells. Over the course of the early 1970s, this cut the price of PV cells to $20 per watt, bringing solar power to something like $2,000/MWh. Much, much better – but still a way to go.

Another turning point for solar power came in 2009, when

Europe set itself the target (at that time wildly ambitious) of getting 20% of its energy from renewable sources by 2020. As a result of this target, countries like Germany, Italy and Spain set themselves renewables targets, and promised producers of renewable electricity a fixed, high price for their power. We've seen (in Chapter 3) that this was terribly costly.[5] The upside was that it worked. With a huge incentivised market, suppliers scaled up their factories and supply chains, driving the cost of the technology down.

Most of this supply response came from China, thanks to a state-sponsored push. Between 2006 and 2013, China's global share of production of PV cells grew from 14% to 60%, in a market that grew eighteen-fold.[6] Prices fell accordingly, sparking a positive feedback loop. Economies of scale appeared through more efficient production lines, better-funded R&D and every other stage of the process. On average, each doubling in capacity has generated a 20% fall in prices (a figure known as the learning rate). By 2019, the cost of panels had fallen to $0.40 per watt.[7] The other thing that improved was the scale of solar projects themselves. The very large GW-scale projects are made up of millions of panels and benefit from increasing automation and digitalisation for installation and maintenance. Utility-scale projects produce electricity at a quarter of the cost of small-scale rooftop panels.[8] So, Europe's policy push did the job, industrialising renewable production worldwide and putting ample and cheap sunlight within reach. If we have a chance to stay within 2 degrees, we owe it to the politicians who started the process fifteen years ago, and to the sacrifices of many unaware consumers.

There are still breakthroughs to come. One fast-improving solar technology uses perovskites, crystal structures first discovered in the Ural Mountains in 1839, which will reduce the cost of converting solar power into electricity. Films much thinner than a human hair can be made inexpensively from solution, allowing them to be easily applied as a coating to buildings, cars or even clothing. Painted on a substrate, or printed using an inkjet-like printer, they make an instant photovoltaic device. Perovskites also work better than silicon on cloudy days and even indoors. If the cost of this new technology falls far enough, solar power generation could unobtrusively move into our cities.

While the story of solar power depended on supply chains scaling up, in wind power what scaled up was the turbine itself – which has become truly massive.

James Blyth, a professor in Glasgow in 1887, was the first to build a windmill that produced electric power. It was a cloth-sailed monster, fully 10 metres high and installed in the garden of his holiday cottage in north-east Scotland. It won him no friends. Blyth offered to light the local high street but the people of Marykirk declared electric power to be 'the work of the devil'. Undeterred, he built another turbine to supply power to the local lunatic asylum.

Wind became a go-to source of power for isolated locations, but the industry took a giant leap forward in 1941, when in Castletown, Vermont, the 1.25 MW Smith–Putnam wind turbine was the first to be connected to a local electrical distribution grid. This brought the promise of much larger-scale development and, as with any technology, scale cuts cost.

Wind energy has its own pioneers, and one the key figures was a civil engineering professor at the University of Massachusetts called Bill Heronemus. In the 1970s, he produced the first detailed plans for wind turbine arrays. The very term 'wind farm' is his, and his and his team's countless technical papers underpin modern turbine technology. He even envisaged grand offshore wind farms where power would be converted to hydrogen, anticipating today's ambitious developments on Dogger Bank and elsewhere in the North Sea.

Dogger Bank Wind Farm takes wind turbine upscaling to a whole new level. Each of its mega-turbines will rise 250 metres from the seabed (making them visible from 55 kilometres away), weigh 2,800 tonnes and provide enough electricity to power 16,000 homes. The scale of these turbines, and of the huge industry behind them, has meant that wind power has become much cheaper.

For an idea of how far renewables have come, just think that solar power at the turn of the century cost \$1,000/MWh. In 2021, a competitive solar project was won in Saudi Arabia at \$10.4/MWh. Wind power cost \$180/MWh in the year 2000; in 2021, an auction was won in Spain for around \$25/MWh.[9] That sort of deflation pales with the cost reductions achieved in the world of electronics, but is almost unthinkable in other areas.

That means that today solar and wind power production is cheaper than fossil fuels in many places in the world. While in Europe coal- and gas-fired power generation comes in at around \$60/MWh, the average cost of electricity from new onshore wind farms in 2020 was as low as \$40/MWh, and the average cost of solar power was \$35/MWh.

Figures like these make renewable energy sources increasingly attractive, and lots of new capacity is being built.

Globally, wind energy almost quintupled to 1,590 TWh in 2020. Solar photovoltaic generation is catching up fast, increasing more than twenty-five-fold in the decade from 2010 to 2020, from 32 TWh to 820 TWh.

Combined, solar and wind power now amount to around 2% of global energy consumption. compared with less than 0.25% a decade ago. While the overall figure is still quite small, the rate of growth is encouraging, and it forms the bedrock of our decarbonisation ambitions. We are no longer in the land of painful trade-offs between economic growth and the environment. We can have our cake and eat it.

Given the success of renewable electricity, a lot of people ask me why we need hydrogen at all. The vision of having a solar panel on our roof and plugging our EVs in at the wall is very exciting, and I will do it myself (as soon as my neighbours agree). But while it is tempting to think that all we need to do is to keep doing what we are doing, just more of it and faster, the truth is that renewable electricity, by itself, will not be enough. It will need a partner.

5

THE UPS AND DOWNS OF
RENEWABLE POWER

**The remarkable rise of renewables is great news, but we
can't simply use clean electricity for everything we do.
Solar and wind power are not constant. Batteries and
other storage systems are limited, especially in the face
of seasonal energy needs. In some industries and long-
range transport, direct electrification just won't work.
Renewables are going to need a partner.**

I am bullish on renewable, and believe we can directly electrify
a lot of our energy consumption, from 20% today to 50, 60 or
even 70% of the final energy mix in a net-zero economy. But
we can't do everything that way. The ephemeral nature of
electricity, the difficulty of storing and transporting it, and the
inconstancy of renewable power – all of these combine to
stymie 100% electrification.

Square peg

The first spanner in the works is that electricity is just not suitable for some applications, what we call the hard-to-abate
sectors. If we can't reach them, we can't reach net zero.

One reason is the difficulty of storing energy as electricity. Lithium-ion batteries have improved a lot – they can be recharged hundreds of times, and they have a higher energy density than other battery technologies – but still they don't hold much energy for a given weight. A kilogram of gasoline holds 13 kilowatt-hours (kWh) of energy; a kilogram of lithium-ion battery holds less than 0.3 kWh. This means electricity isn't going to be the best way of powering sectors which need to take lots of energy with them. Just think of long-haul flying: we would need so many heavy batteries that the plane wouldn't be able to take off. And batteries won't compete with molecules to propel cargo ships across oceans. While for shorter distances battery-powered trucks look like a hot contender, for long-distance trucking, the amount of space and weight that the batteries would add (and the difficulty of charging them quickly) makes it hard to imagine their widespread use.

Next up is heavy industry, such as steelmaking, where electrification is difficult to impossible. This is not due to immature technology, or any lack of political or commercial will. It is due to basic physics. If you are in the chemical industry, you might be using fossil fuels to synthesise your products. You can't magically make molecules out of electrical current instead. Other industries need intense heat, which is expensive to provide with electricity. You can use it indirectly to heat a medium (usually air) that surrounds the object you want to heat, which works well enough for roast chicken, but at high temperatures this indirect approach is inefficient. Other methods, such as microwaves and induction heating, heat the product directly – but it's hard to attain very high temperatures.

Overall, hard-to-abate sectors – including heavy road

transport, heavy industry, shipping and aviation – account for around a third of energy-related emissions today.

Long-distance sunlight

Another issue is that electricity can be hard to transport over very long distances.

In all essentials, the power grid we use today is one we've inherited from the previous century. High-voltage alternating current (HVAC) is transmitted long distances along cables that are kept well out of the way of people, either on pylons, or (less frequently, because of the cost) underground. At substations in each neighbourhood, this electricity is stepped down on to smaller, lower-voltage power lines. Finally, transformers step the voltage down to the level that will operate domestic equipment without setting it alight.

Up to 8% of generated electricity is lost between the power station and your home. You can actually hear this happening if you're close to a high-voltage power line – it's that fizzing, crackling sound. You can also see it happening: as the metal cables heat up they expand, and begin to sag in the middle. That accounts for about a quarter of the energy lost. The other three quarters are lost, rather less spectacularly in the lower-voltage lines to our homes.[1]

These losses have never seemed too onerous, because we've learned not to transport electricity very far. In Italy, we are on average 25 km away from our nearest power station. By contrast, we are something like 1,000 km away from our main natural gas supplier, because it takes little energy to push gas through a pipeline. The same goes for coal and oil, which often travel thousands of kilometres on ships.

In the future, however, our hunger for renewable power means we will need to look further and further afield for our supply – especially if locally there are land constraint issues, or difficulties in reaching the required scale in renewable production in the required timescale. We are also going to want to access the cheapest places to produce renewables. That makes importing renewables an interesting idea. The question is how best to do so. High-voltage alternating current, as we've seen, loses energy along the way. There is also the possibility of using high-voltage direct current cables, which have lower energy losses and may work better for long distances. Both of these are expensive ways to transport energy compared to the current costs of moving it as natural gas in pipelines, but in many cases they may be the best option to get renewables into your car or home lighting system. In many others, however, they will not be.

Overall, the difficulty in long-distance electricity transport means that we generally prefer to put our wind farms and solar panels close to where the power is needed – within a few hundred kilometres, in any case. And that can be both impractical and wildly inefficient. It's not very sunny in northern Europe, so a solar panel in Stockholm produces a lot less juice than the same panel sitting in the Sahara. Trying to decarbonise the world with hard-to-transport renewables would also create lots of separate energy markets, with different energy costs. And that would make life very difficult for countries facing relatively high costs to compete in a global market.

Easy come, easy go

Green electricity production is intermittent, depending on the whim of sun and wind. The sun doesn't shine at night. There are weeks when the wind dies and the leaves barely rustle in the trees. There are times, called *Dunkelflaute* in German, when it is both dark and calm, so there is hardly any renewable power available for use. Sometimes this happens when the weather is cold and so demand for heating is high: the dreaded 'cold *Dunkelflaute*'. Strong winds can generate more wind power – but when they reach 90 kilometres per hour production falls off abruptly, because many turbines shut down for safety and to minimise wear.

Intermittency is a challenge because the grid must always balance demand and supply. That's to do with the nature of electricity. You only get an electrical current in a wire when you give the electrons a push with a generator or a battery or some other device that creates a voltage between the two ends of the wire. Almost as soon as you stop pushing, the current stops. So one hungry customer's sudden demand for electricity has to be met more or less instantly, or else a brownout – a serious reduction in voltage – hits every customer fed by the same power line.* That contrasts with a gas network. When you put natural gas in pipelines, you can vary the pressure so as to pack it in more or less tightly, meaning that you have lots of flexibility to satisfy demand fluctuations without resorting to additional generation or storage. Keeping the electricity system in balance is akin to tightrope walking, while balancing the gas grid is a walk in the park.

How can the power grid address imbalances between supply and demand?

---

* According to a calculation by engineering firm Siemens, if one in ten of us turned our washing machines on at the same moment, we would bring civilisation to its knees, for at least a few hours.

On very short timescales, balancing the grid is a problem that partly takes care of itself. Power stations employ big, heavy turbines to generate electricity, spinning at round 3,600 times a minute. All this fast-rotating heavy metal conveniently works as a gigantic flywheel, storing energy and stabilising the grid for up to about, well . . . fifteen seconds.

For anything longer than that, other sources can step in to take up the slack. This is known as dispatchable power. Gas-fired power stations can be ramped up in a matter of minutes to meet demand. If you are lucky enough to have a lot of flexible hydropower – a renewable source that is also on-demand – like Norway or the Alps do, you don't need much dispatchable gas; elsewhere it has become vital.

This isn't ideal, though. Having gas and other fossil fuel power plants constantly ramping up and down costs money and generates emissions. Power plants must often operate at partial capacity, instead of at their most efficient setting. If these back up plants don't run very often, their owners might be tempted to close them. But that won't do, because although these facilities may not often provide electricity, they provide an essential service, stepping in when we need them; without them we would face costly blackouts. So it usually falls to the system, and therefore ultimately to consumers, to pay gas-fired power plants to just hang about in case they are needed.

Having lots of renewables producing at once, especially if combined with a dip in demand, can cause negative prices, when those who generate electricity have to pay their customers to take it. We've just seen lots of days like that during the 2020 Covid-19 lockdowns, when energy demand from business declined dramatically, far outpacing increased consumption in homes. There were roughly 300 hours (almost two entire weeks) of negative prices in Germany alone. Clearly, this is a problem for producers, rather than customers. But that turns into a problem for the system as a whole, because if investors in electricity generation struggle to make a return, there won't be many following in their footsteps. Loss-making production eventually translates into constrained supply.

Of course, the opposite scenario is far more worrying. What happens when demand for power suddenly rises on a still, grey day? Today, the network operator can ask a gas-fired power station to take up the slack, but already there are times when that is being tested to the limits. In January 2021, a UK power shortage meant that West Burton B, a gas-fired power station in Nottinghamshire, sold its output for £4,000/MWh, a hundred times the normal wholesale price. UK balancing costs, which are all the costs paid by the system operator to wind and gas companies to change their production levels to safeguard the system, were £1.8 billion in 2020.[2]

Extreme weather, expected to increase as a result of climate change, can cause particular problems. We saw a perfect storm of demand and supply in the Texas polar vortex of February 2021. Freezing temperatures not seen in more than thirty years led to a huge spike in energy demand. At the same time, wind turbines were pushed offline because of the cold, and 40% of fossil fuel capacity was unavailable because either the power plants froze or gas production facilities did. Power prices skyrocketed, reaching the administrative cap of $9,000/MWh. A series of outages left 34% of all customers (4.3 million homes) with limited or no power.

While not primarily caused by the intermittency of renewables, the Texas disaster underscores the fragility of power grids – something I witnessed first-hand when I worked for Italian utility Enel, at that time owner of electricity system operator Terna. On 28 September 2003, I woke up to find fifteen unanswered calls on my phone. A few hours earlier, a cable importing power into Italy had come into contact with a nearby tree. Suddenly, the whole of Italy was in a blackout. As hour after hour went by, we started to get distress signals from essential facilities such as hospitals, which really brought home to me just how much we rely on electricity, and how important it is not to put all one's eggs in one basket. If a tree hits the line, we don't want the guys sent to sort it out driving an electric vehicle. On the day of the Italian blackout, one of our concerns was whether it would affect operations on the gas system. Thankfully, the answer was no. My counterpart at Eni (at that time the owner of Snam, which runs the gas transport system) assured me that the gas grid had a totally analogue system as a backup.

Blackouts and brownouts are a more frequent problem in the developing world, where the World Bank estimates they cost over $150 billion per year.[3] Meanwhile, the impact of these events is only going to increase going forward, in an increasingly hyper-connected world. Like it or not, we're now reliant on connected data and communications systems, which make us vulnerable to even short interruptions in supply.

So intermittent solar and wind power already bring challenges, even when they provide only a limited share of our electricity and we can call on fossil fuel plants as backup. Today, in Europe around 32% of our electricity comes from renewables (and we use electricity for around 20% of our energy consumption overall). What will happen when we fully decarbonise and can't use fossils to solve intermittency? We will need to change the way we transport, store and use electricity – and it will still not be enough.

We can start by extending our power grids. It will probably be sunny or windy somewhere, and times of peak demand for one place may be times of slack somewhere else, so a big well-connected grid will tend to balance better. As John Pettigrew, CEO of the UK's National Grid has put it 'the bigger the network, the more interconnected, the more stable it is'.[4]

Even then, we will need to store a lot of energy. The traditional solution here is water. At times of low demand, spare electricity is used to pump water from a lower reservoir to a higher one; then when we want extra power, we let it run from the higher reservoir to spin a turbine. Pumped hydro systems are responsive – they'll be generating power within fifteen seconds of us pushing the button – and they're between 70 and 80% efficient. Pumped hydro accounted for around 95% of global power storage capacity in 2020 – but that's still not saying much. Pumped hydro projects now store up to 9 TWh of electricity globally, only 0.03% of total power generation (which amounts to 27,000 TWh).

The lower reservoir doesn't have to be anywhere special. It could even be the ocean. In 1999, Japan had a unit in Okinawa using the sea as a lower reservoir. In the Netherlands, meanwhile, there is a clever plan to use the sea as the *upper* reservoir – a consequence of that country being in large part below sea level.

But most pumped hydro schemes need mountain lakes to site their upper reservoir, and such places are scarce. Smaller still is the number of such places you would want to build on: pumped hydro and areas of outstanding natural beauty are not a good mix. And, even if companies were to get permission and acceptance from local communities, who typically rely on the same water or land for their own benefit, construction is not particularly cheap.

The other obvious option is batteries. The main issue with battery storage is how expensive the battery itself is, and how much use you are going to get out of it.

The cost of lithium-ion technology has fallen to around $120/MWh in 2020, which is much better than it was, but still around twenty-five times the cost of storing natural gas in underground reservoirs ($6/MWh). That adds a hefty extra cost on top of solar power production, which as you may remember can be as low as $10.4/MWh, and on average in Europe is around $35/MWh. It is known as an 'integration cost', in that it reflects the cost of addressing the intermittency of the power which is produced in this way. For solar and wind, integration costs are often added to the production cost to give a true comparison with power produced from fossil fuels, which is more expensive to produce but doesn't require all these additional batteries because you can turn it on and off at will. Integration costs rise with the percentage of renewable power generation, and can reach astronomical levels when you are trying to get from a grid made up of 80% solar and wind to one that is 100%.

At these prices, the solar plus battery combination is more expensive than gas-fired power generation (say $185/MWh vs $60/MWh), but it is likely the best 'zero carbon' option. And, indeed, some are already being built. In 2017, after a series of severe storms forced grid closures, Tesla won a bid to build a 100 MW, 129 MWh battery storage to stabilise the South Australian power grid. Southern California Edison is set to inaugurate a 100 MW, 400 MWh unit in Long Beach, California, in 2021.

However, a cost of electricity storage around $120/MWh assumes we are using the battery pretty much daily (and so spreads the cost of the battery over a lot of electricity). The less you use it, the more the cost of storage per MWh increases. Batteries have other shortcomings. They are made out of metals which are, today, extracted and processed in a limited number of locations. On top of this, lithium mining requires large amounts of water in often arid locations, and can lead to contamination of soils and rivers. And batteries are difficult to recycle well, so they may end up in landfill instead of being recycled at all, leading to more contamination.

It would help if we could plug in a billion free batteries. That's the number of cars on the road, most of which are likely to be electric eventually. Could we store spare power in the batteries of these cars? UK company Octopus Energy, which has its tentacles in a lot of innovative clean energy projects has been running recent vehicle-to-grid trials in which participants lease a brand-new Nissan Leaf and have their homes kitted out with a special charger. This allows their car to both take electricity from the grid when there isn't a lot of demand, and sell it back when supply is tight, using an app developed by Octopus and a company which is today called Engie EV Solutions, at the time chaired by climate expert and author Chris Goodall. This is still an early-stage idea, but well worth investigating further.

Innovative forms of energy storage may also step in. Where a flat landscape doesn't allow pumped hydro, we could use compressed air, pumping it into a cavern to store energy; or hydraulic pumped storage, lifting a heavy block with water pressure. You can also leverage the force of gravity by using excess power to lift something other than water: Swiss start-up

Energy Vault, whose project won the Eni Award for World Changing Ideas, has designed a six-armed crane to lift and stack cement blocks when power is ample, and then lower the blocks when power is needed, generating electricity through their downward motion. Power can be stored as heat in water tanks, bedrock or molten salt. And for the summer cooling side of things, air conditioning systems can effectively store energy by making ice overnight and using it to cool buildings during the day. But these too are immature technologies.

We can also enlist the help of consumers through demand response. People would delay some energy uses, such as washing clothes, when there is scarce supply, waiting until the grid has lots of potential renewables at hand. Given inertia is a very human trait, I wouldn't bet on customers changing their behaviour spontaneously, without the help of technology. We need transparent energy pricing and smart meters to pick the best time to turn things on and off. But while this is likely to be a useful piece of the puzzle, it won't change the fact that we are going to need vast amounts of electricity storage to smooth intermittency, and that these will add to the cost of renewables.

Even if we extend our grids and build all of this storage, it still won't be enough. We will still need dispatchable power to cope with rare events such as the Texas storm of 2021. Giant banks of batteries, sitting idle till they are needed for such emergency backup, would be prohibitively expensive.

Cold comfort

Nothing we've talked about so far is likely to give us the astronomical numbers and storage characteristics needed for winter heating. Two opposing trends combine in a winter crunch. In winter it isn't very sunny: at European latitudes, solar radiation in the summer is two to five times that in winter (although wind fares better). And we tend to use a lot more energy, mainly for heating. In many temperate zones there's about a threefold difference in fuel consumption between summer and winter.

Similarly to intermittency, this isn't usually a problem today because the seasonal swings in renewable power productivity are smoothed over with fossil fuel power generation, and the seasonal swings in energy demand are supplied by the gas system, which can store energy in large amounts underground, and transport it through its grid.

But seasonality is a huge hurdle to full electrification. For the colossal amounts of energy needed to balance the grid across seasons, batteries are no use. They would be ruinously expensive. Europe consumes about 5,300 TWh each year for heating and cooling, most of it in the winter.

As we saw when looking at intermittency, the cost of storing energy in batteries is something like $120/MWh if you are charging and discharging over 300 times a year. If you want to store energy in July and release it in December, that would only be one cycle a year, which means the cost of electricity coming out of the battery could be hundreds of times the quoted figure. And if batteries are a no-go, so is pumped storage. There aren't enough suitable mountain lakes to cover our seasonal swings.

What if we were to try to do away with storage altogether, and meet winter demand peaks through additional renewable production capacity? Well, that would mean massive overproduction of renewables in the summer and would be a waste of money and land.

Seasonal demand would also overwhelm today's power grids. In Europe, energy demand peaks during the winter, when temperatures drop. Even in places that are not extremely cold, such the UK, winter peak energy demand is several times the capacity of the electricity network. At the moment, all this extra energy is carried by the natural gas network, which has been designed to cope with it, and ensure supply when required. Indeed, in Europe, peak demand on the gas grid in winter is five times electricity demand: 2,500 GW for gas against 590 GW for electricity.[5] If all energy had to be delivered as electricity, that would require a huge upgrade of the grid. We would also waste a perfectly good gas transport asset, which consumers have already paid for.

For me, that extinguishes any thought of fully electrified heating. Not everywhere, of course. Places like Australia and California can rely on fairly steady weather through the year, with no drastic changes in the supply of renewable electricity or the demand for energy. Full electrification would work well in such circumstances. But the three places I've lived in – the east coast of the US, the UK and Northern Italy – are less balmy and will need a form of storage that can swallow as much spare energy as we care to throw at it.

The vision of an economy powered entirely and directly by green electricity has taken a fair number of knocks in the last few years, with net zero targets forcing us to think through all of the challenges of full electrification. That's not to say that the revolution in renewable power isn't a huge step forward. It is, and it creates the conditions for full decarbonisation. We just need something else, alongside electricity.

Something that takes the spare solar power from the summer

and stores it for winter, and provides dispatchable generation when needed. Something that can be transported like gas, linking regions and seasons, and allowing us to tap the desert sun and mid-ocean wind, wherever renewable power is most efficient and we have the most space to put it. Something that can be delivered through existing, robust gas infrastructure, reducing the need for massive new investments in the electricity grid. Something that can reach those hard-to-decarbonise sectors, which require different forms of energy.

We need a *molecule*.

By contrast with electrons in a wire, molecules can store energy, even for millions of years, in a liquid, like oil, or a solid like a lump of wood or piece of coal, or even a gas. When called into action they can provide power at a second's notice. This makes molecules ideal for long-distance travel, long seasonal storage and management of the increasing intermittency of supply and demand. But of course, we can't use oil, gas or coal in a fully decarbonised world.

We need the *right* molecule.

The missing link in the decarbonisation story is a crossover energy vector; something that straddles the worlds of electrons and molecules, able to weave together different strands of the energy system, deliver ample clean energy to power a growing global population, and level out the cost of renewables across regions, creating a fairer global economy and a fairer transition.

Let's take a closer look at the simple molecule that can serve as this missing link.

Part 2

Hydrogen: A 'How To' Guide

6

ENTER HYDROGEN

Hydrogen, the simplest and most abundant element in the universe, has long held promise as a carrier of energy. Early experiments revealed its great potential power and its vital, intimate relationship with electricity. While both these traits sparked enthusiasm in scientific circles, hydrogen's promise as a great energy connector remained unfulfilled.

Hydrogen's story goes back 13.7 billion years, to a time when the universe was new-born, and very hot. For the first 370,000 years, space was filled by a hot particle soup known as a plasma, made up of loose electrons and protons (plus a few heavier nuclei – combinations of protons and neutrons). Eventually temperatures dropped to the point where electrons could bind with protons to form hydrogen atoms. Hydrogen emerged from that primordial furnace in far larger quantities than any other element, and even today it dominates the cosmos.* It is the main ingredient of stars,[1] including over 90% of our sun, and a thin mist of it is scattered through space. Sometimes, in giant interstellar gas clouds, it

---

* If we ignore the mysterious phenomena that astronomers call dark matter and dark energy, which don't form into interesting things like planets and people.

adopts the form we usually see on Earth: two atoms joined together to make a hydrogen molecule, H_2.

Hydrogen is the simplest atom: just a single proton orbited by a single electron. This simple structure is the source of all its properties, both wondrous and troublesome. This primal element's promise as a great connector of energy began to emerge long before we knew its origin or inner nature.

We get our first glimpse of hydrogen in the sixteenth century, from the experiments of Theophrastus Bombastus von Hohenheim, a Swiss physician whose talent for self-promotion was so great, the very word 'bombastic' is derived from his name. Better known to us by his pseudonym Paracelsus, he found that he could dissolve iron in sulphuric acid, and a mysterious gas was given off.* His follower Théodore Turquet de Mayerne, repeated the experiment and set this mysterious gas alight, discovering just how flammable it was.

Jump forward to 1766 when, in his private laboratory in London, Henry Cavendish collected bubbles of the gas (generated through a similar reaction to Paracelsus and de Mayerne but using hydrochloric acid and zinc). For a while he took a simple delight in setting these bubbles alight. He did notice, however, that burning his gas generated an unexpected by-product: water. By 1781, he had established what we all know today: that water is, in fact, a combination of two gases. Antoine Lavoisier, a French nobleman whose scientific genius was snuffed out in the French Revolution, gave these gases their modern names, hydrogen (*water-forming*) and oxygen and ushered in the modern age of chemistry.

* As we now know, iron reacted with the sulphuric acid to make iron sulphide and molecular hydrogen gas: $Fe + H_2SO_4 \rightarrow FeSO_4 + H_2$.

Now, with the hindsight of physics, we have a deeper under-standing: hydrogen's lone electron is easily captured by other elements to form new substances, such as water. In those explo-sions that so entertained Cavendish, hydrogen bonds with oxygen atoms to form H_2O, while releasing a lot of energy.

De Mayerne, Cavendish and Lavoisier were all enchanted by hydrogen's flammability, which already hinted at its enormous energetic potential. This was confirmed by Lavoisier and the polymath Pierre-Simon Laplace, who measured how hot hydrogen got when they set it alight. The results of their experiments were off-the-scale. It turns out that burning 1 kilogram of hydrogen releases enough energy to drive a typical car for 130 km or provide two days of heating for an average household.

Soon we saw an inkling of hydrogen's intimate relationship with electricity, which is at the heart of today's vision for a green future. One Sunday in 1792, by the bank of Lake Como, the inventor Alessandro Volta generated an electrical current between two metal plates that were separated by paper or cloth soaked either in salt water or sodium hydrox-ide. This 'voltaic pile' was the first battery.* Just six weeks

* Volta's enlightening discovery had quite gothic origins. For more than a decade, from 1780 to the early 1790s, it seemed to researchers that animal life was driven by a newly discovered life source, dubbed 'animal electricity'. This was a notion cooked up by the Bologna-born physician Luigi Galvani, to explain a discovery he had made in 1780 with his wife Lucia. They had found that the muscles of dead frogs' legs twitch when struck by an electrical spark. Galvani concluded that living animals possessed their own kind of electricity. The distinction between animal electricity and metallic electricity didn't hold for long. By placing disks of different metals on his tongue, and feeling the jolt, Volta showed that electricity flows between two metals through biological tissue.

after Volta's announcement, two British scientists, William Nicholson and Anthony Carlisle performed an experiment that, while less often covered in the history books, proved no less vital. In 1800, the pair set about recreating Volta's setup, attaching wires to either side of a voltaic pile before dipping them into a container of water. Bubbles appeared on the submerged wires: an electric current was splitting water into its constituent gases. They had invented the electrolyser – the device that now enables us to generate hydrogen using renewable power.

The first really solid explanation of electrolysis came from German chemist Johann Wilhelm Ritter, an independent and lively mind, Ritter counted both Goethe and the polymath Alexander von Humboldt among his friends.

Here's a simplified recipe for Ritter's set-up. Take a vessel, fill it with water, and dip two metal bars part-way into the water. Connect the dry parts of the bars to a battery and, hey presto, you've created a couple of electrodes. The voltage from the battery drives chemical reactions at each electrode. At the one side, water molecules split up, generating oxygen gas, loose protons and loose electrons. The electrons are pulled in by the positive electrode, the oxygen forms a gas, and the protons wander off through the liquid. These protons are given electrons at the negative electrode, producing a gas of hydrogen molecules. Then, invert two glass vessels full of water over each electrode and watch as bubbles rising from each displace the water in each vessel, filling them with gas.

A basic feature of many electrolysers, which Ritter felt he didn't need, or in any event didn't come up with, is a diaphragm which, while it won't interrupt the current flowing through the water, does prevent bubbles of oxygen and bubbles of hydrogen from meeting and reacting (for which read: exploding).

One last detail: pure water does not conduct electricity well. So, we need to add an electrolyte: another chemical to get our electrolysis going. Salt will do the job, or small amounts of sulphuric acid, along with many other electrolytes that are used to improve the speed of different kinds of real-world electrolysers.

Not long after Nicholson and Carlisle had invented the electrolyser, attempts were made to stand the thing on its head. The idea was that the electrolyser ought to work just as well in reverse, as a device we now call a fuel cell. Hydrogen atoms enter at the anode where a chemical reaction strips them of their electrons. The positively charged protons pass through a membrane to the cathode, and the negatively charged electrons flow through a circuit to meet them. Finally, the electrons combine with the protons and oxygen from the air to make the fuel cell's by-products: water and heat.

Credit for the first fuel cell is split between a German-Swiss chemist, Christian Friedrich Schönbein, and a Welsh judge, Sir William Grove, who both made the same discoveries, via much the same experiments. Grove's second cell, invented in 1839, was the forerunner of modern fuel cells. He took two platinum electrodes and dipped them in a container of sulphuric acid. The other end of each electrode he sealed in separate container, one holding oxygen, the other hydrogen. A current immediately flowed between the two electrodes.[2]

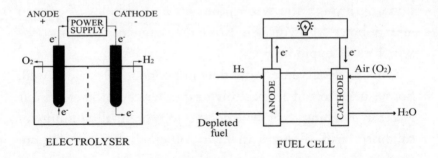

ELECTROLYSER   FUEL CELL

Electrolyser and Fuel Cell

To many writers, hydrogen instantly became the fuel of the future. In the 1874 novel *The Mysterious Island*, Jules Verne imagined that water, decomposed by electricity into its primitive elements, would 'one day be employed as fuel, that the hydrogen and oxygen which constitute it . . . will furnish an inexhaustible source of heat and light'.

Soon after this, one visionary seemed to be fulfilling Jules Verne's dream. Poul la Cour was a Danish inventor who worked in telegraphy in the 1870s before turning his attention to education. He was particularly concerned with the plight of young people growing up in the countryside, even as heavy industries were emerging and cities were expanding. La Cour saw that to survive, rural communities needed to modernise. And for that, they had to get access to two things that cities enjoyed in abundance: education, and plentiful energy.

La Cour's genius was to address both needs at the same time. Through his teaching, and by some enlightened political jockeying, he fostered a generation of regional engineers. His idea was to make the Danish countryside a self-sustaining enterprise, independent of the big cities.

La Cour's first love was wind power. If Denmark was ever to compete with its neighbours, as the Industrial Revolution swept across Europe, then it needed a reliable source of energy. The country lacked coal, but it had plenty of wind. Such an early interest in wind power was visionary, for sure, but it didn't come out of nowhere. In the Netherlands, the idea of electrification by means of windmills had already been investigated. The attempt was abandoned because of two seemingly overwhelming difficulties. First, traditional Dutch windmills were hopelessly inefficient at generating electricity, and no one could see an obvious way to improve them. Second, once electricity was generated, it had to be used straight away, because when the wind died, so did the electrical supply. Some means of electrical storage was needed, but batteries at the turn of the century were impossibly expensive.

La Cour revisited both challenges. He redesigned the classic windmill, using redesigned sails to turn a dynamo to generate electricity.* To address the second challenge, how to store the electricity he had generated, la Cour converted an old water mill near the town of Askov into a windmill and used the electricity he generated to produce hydrogen by electrolysis. In collaboration with the Italian physicist Pompeo Garuti, la Cour filled tanks with hydrogen and oxygen, and used the hydrogen directly as a fuel. This was no kitchen-table effort, either: production reached 1,000 litres of hydrogen an hour.

La Cour's mill lit Askov's Folk High School, where he taught, from 1895 until 1902, and the place suffered not a single day

* His government backers, expecting something outlandish, were disconcerted to discover that la Cour's devices still looked like, well, *windmills*, and they threatened to cut off his funding. Happily, the efficiency achieved by la Cour's technology persuaded them to reconsider.

without light thanks to the 12 cubic metres of hydrogen stored in the mill's hydrogen tank. In 1902, the windmill in Askov became a prototype electrical power plant serving the whole village, an arrangement that held until 1958, when batteries and a petrol engine (for reserve power) took over.

So hydrogen's power was being demonstrated more than a century ago. We knew then that a kilo of the stuff could hold a tremendous amount of energy. We already had the basic tools to turn electricity into hydrogen and back again, on demand. But while at least one village in Denmark was enjoying some hydrogen power as early as 1902, it did not catch on around the world. Why not?

Two fundamental problems scuppered early attempts to harness hydrogen. Its low density made it hard to handle. And in contrast with handy, abundant fossil fuels, it is frustratingly difficult to isolate from other elements on Earth.

7

FLIGHT OF FANCY:
HYDROGEN'S EARLY USES

With a single proton orbited by a single electron, hydrogen is not only the simplest but also the lightest element in the universe. This makes it more difficult to store and handle than dense fossil fuels. For a long time we turned away from its energetic aspects, and instead made use of this unparalleled lightness to lift us into the skies. But that wasn't hydrogen's true calling.

One cubic metre of hydrogen weighs just 89 grams.* It occupies 3,000 times as much space as gasoline, for equal amounts of energy. So hard was it to corral hydrogen that, for well over a century, we ignored any applications of its energetic potential and instead concentrated on exploiting its lightness.

The Montgolfier brothers went aloft in the first hot-air balloon in the summer of 1783. No one, at this time, was quite sure why their balloon managed to rise into the sky. In as much as they ever expressed an opinion, the Montgolfier brothers argued that it was the smoke from wet hay that acted as a lifting agent. The brothers were gifted inventors, not scientists.

---

* At standard atmospheric pressure and temperature.

Antoine Lavoisier – who was as taken up with balloon mania as everyone else of his generation – realised that hydrogen would be far lighter than hot air, and put pen to paper with a vision of balloons lifted by extraordinarily light hydrogen gas. This was before the invention of the electrolyser, so Lavoisier needed to find his own way to split water. And in the winter of 1783–4, he found one. Collaborating with the army officer Jean Baptiste Meusnier, Lavoisier worked out how to generate hydrogen by passing steam through the red-hot barrel of an iron cannon.

Jean-François Pilâtre de Rozier, a physics and chemistry teacher, had already flown with the Montgolfiers when he caught the hydrogen bug. Controlling a balloon's altitude was crucial, but as yet a rather haphazard affair. Pilâtre de Rozier had the bright idea of using a combination balloon, in which an outer hydrogen layer provided most of the lift, while a hot-air core controlled the altitude of the flight.

Convinced that their design would revolutionise ballooning, Pilâtre de Rozier and his companion Pierre Romain set off from Boulogne-sur-Mer on 15 June 1785 on an attempted Channel crossing. After about half an hour, and with the wind obstinately forcing the balloon back to shore, the two, showing 'signs of alarm', fought to lower the grating over their brazier – but it was too late. 'The inflammable contents of the larger sphere soon filled the vacant portions of the silk and pouring down the tube which formed the neck of his balloon speedily reached the furnace, which was disposed at its lower extremity, and became ignited.'[1]

The bang was an almighty one, and the design for a double balloon were quietly shelved (only to be dusted off when the inert lighter-than-air gas helium became available). Otherwise, and incredible as it seems now, no one was particularly put off by this accident. Yes, hydrogen was flammable, but so what? Hot air balloons, lifted by straw burning in a brazier hung above a wicker basket, were hardly less perilous. And so for 150 years more, inventors and pioneers insisted on using this explosive gas as a buoyancy aid.

Ferdinand von Zeppelin, a retired German officer, resigned from the army in 1891 and set about creating a flying weapon, lighter than air, filled with hydrogen, and held together by a steel framework. Zeppelins, which had started life in Count Ferdinand's notebook as mail carriers, were to terrorise Western Europe throughout the First World War, travelling at about 85 mph and carrying 2 tons of bombs. Nor were they particularly easy to knock out of the sky. Hydrogen is so light, it disperses very quickly. An ordinary bullet, while it might puncture a Zeppelin's gas bag, stood no chance of igniting the hydrogen. Special explosive bullets had to be developed to clear British skies of the airborne threat.

Following the war, the Zeppelin design found a role in thrilling peacetime projects, from polar explorations to circumnavigations of the globe. Airships like the LZ-129 *Hindenburg* and her sistership LZ-130 *Graf Zeppelin II* established regular commercial air travel between continents.

Safety was their big selling point: the *Graf Zeppelin* flew for more than 1.6 million kilometres without a single fatality – a record no aeroplane of the time could match. But they handled bad weather poorly, and the number of weather-related airship accidents was already mounting when, on 6 May 1937, the *Hindenburg* docked at a mooring mast at Naval Air Station Lakehurst in New Jersey, burst into flame, and crashed to the ground. Of the ninety-seven people on board, thirty-three jumped or fell out of the airship and two more died of burns from the fabric and diesel fuel. One member of the ground crew was killed when a motor fell on him.

Whether the *Hindenberg*'s hydrogen supply caught light or not remains controversial, and is somewhat beside the point: it certainly failed to keep the craft in the air. Terrifying movie footage of the incident, and a vivid, heart-rending radio broadcast live from the site, brought the era of passenger airships to a swift end. It also created a strong association in people's minds between hydrogen and raging infernos, a largely unwarranted perception of danger that we will return to in Chapter 20.

Ballooning was an ultimately unsuccessful detour for hydrogen, whose main selling point is its vast energetic potential rather than its lightness. Yet it took us so long to make use of this, largely because hydrogen was outcompeted by abundant and cheap fossil fuels.

8

THE ALLURE OF OIL

On Earth, hydrogen is locked away in other molecules, while fossil fuels are freely available. That made it difficult for it to compete with fossil fuels. But the two weren't competing on a level playing field. There are hidden costs to coal, oil and gas that were not factored into the equation. Properly accounting for them highlights how important it is to free hydrogen from its molecular shackles.

Apart from its cumbersome nature, the other main reason that hydrogen has struggled is that you can't just dig it up and burn it. You rarely find it in pure form on Earth[1] because there are lots of other elements around, and hydrogen eagerly grabs on to them with its reactive lone electron. So we usually find it chemically combined with something else.

Much of it is bound up in water (H_2O), which covers 75% of the globe's surface. Sixty per cent of the atoms in our bodies are hydrogen. It is present alongside carbon in most organic compounds like food (such as carbohydrates – it's in the name) and fuels made from old organic matter (hydrocarbons, made from hydrogen and carbon in varying quantities). Both of these elements, when they combine with oxygen, generate energy. But the carbon also generates carbon dioxide (CO_2), the hydrogen only water (H_2O). So, the cleaner fuels are those with more hydrogen.

Table 4: Hydrogen content in different fuels

	Carbon content	Hydrogen content	CO_2 emissions per MWh produced [kg]
Coal	>90%	5%	900
Crude oil	84–87%	11–13%	565
Natural gas	75%	25%	365
Hydrogen	0	100%	0

Hydrogen also likes to bond with itself. When it is generated in a chemical reaction, it generally doesn't come as a single atom, but as a pair of atoms clinging together, H_2. This is the lightest molecule in existence, and because of that, it moves fast. In the upper reaches of the atmosphere, a hydrogen molecule can reach escape velocity and shoot off into space, which means there is essentially no hydrogen in our atmosphere either.

So getting hold of hydrogen means breaking it free of its bonds with other atoms, which takes at least as much energy as you get back when you burn it. For a while, that was enough to relegate hydrogen to the fringes of the energy market. What was the point of us using up electricity to free hydrogen from its watery prison if we could use electricity directly? And where you couldn't use electricity directly but needed a portable, dense fuel, there were plenty of ready-made ones lying around in the form of fossil fuels. These are the product of plants and animals that worked millions of years ago to store the sun's energy in their bodies. That organic matter was buried and slowly transformed by the heat and pressure into coal, oil and natural gas. It's hardly a surprise that humans took advantage of this. Apart from the effort involved to extract them, fossils seemed to be a free resource.

But this doesn't account for two things.

First, fossils are limited; once we use them, they are gone. So by burning them with abandon we were in effect selling off the family silver, rather than setting up a sustainable energy system. To get an idea of how precious a resource oil is, consider that it takes nearly 25 tonnes of prehistoric buried plant material to make a litre of petrol.[2] A tonne of poplar wood today is €90. So to get the 25 tonnes you'd need to make a litre of petrol, you would have to shell out €2,250. And then wait for a few million years while the plant material bakes at high pressure. In this context, petrol at the pump is wildly undervalued. Or, to put it another way, we are using up plants that would have required 40 acres of land every time we go visit grandma 20 miles away. These estimates are from Jeffrey S. Dukes, then an ecologist at Purdue University, who said: 'Every day people use the fossil fuel equivalent of all the plant matter that grows on land and the oceans over the course of a whole year'.

And of course, as we now know, digging up old carbon and pumping it into the atmosphere is not a very clever thing to do. The environmental cost of burning fossil fuels is high, whichever way you look at it. One way of determining the price of CO_2 is to think about what extra cost you would need to impose on fossil fuels to make the least expensive clean alternative convenient. By that reckoning, you are looking at around US \$30/tonne of CO_2 today (which adds \$13 to the price of a barrel of oil). That's quite low because we are still at the beginning of our decarbonisation journey, and there are still lots of low-hanging fruits. As we progress to harder sectors to decarbonise, the additional cost of CO_2 that would need to be imposed on fossil fuels to encourage switching would rise to around \$120–130/tonne[3] by 2050. Another way of thinking about what CO_2 should cost is to look at how much it would cost to capture it before or after it is emitted. Today, the cost of carbon capture and storage (CCS) is around \$90/tonne for a baseload gas power plant,[4] and the cost of capturing CO_2 from the air (DAC) is around \$200/tonne,[5] although that could fall to \$60/tonne after 2040.[6] At that price, it may be cheaper to remove some CO_2 from the air through DAC than to do other things which are higher up the abatement cost curve, making it a really useful tool in our toolbox.

Of course, the cost of carbon will be even higher if we *don't* abate it. Consider the cost impact of global warming on agriculture, fisheries and timber, through ecosystem services such as flood protection, right up to the value we put on other species and the natural world. In the most extreme scenarios, this last value is arguably infinite. A 2020 model of the ecological costs of emissions put the total cost at \$160 per tonne of CO_2; meaning our present emissions are costing more than \$6 trillion per year, or about 8% of global gross domestic product (GDP). Pricing this in, the study concludes that our most cost-effective pathway is to hit global net zero by 2050 and stay under 1.5 degrees.[7]

These costs were for many years ignored, which is what led us to choose coal, oil and later natural gas as our preferred energy sources.

Where we do use hydrogen today (mainly in non-energy uses, such as feedstock to make fertilisers or for use in the process of refining oil) we make it from fossil fuels. 16% of the hydrogen we make today comes from coal. Another 30% comes from oil, a cleaner source, but not by much. Fully half of our hydrogen comes from natural gas (mostly methane, CH_4) from a process called steam reforming. In this process, natural gas is mixed with steam, then passed at high temperature and pressure over a nickel catalyst, which splits the gas–steam mix into hydrogen, carbon monoxide, carbon dioxide and water. More steam is added, and another catalytic converter turns the carbon monoxide to carbon dioxide, leaving you with water, hydrogen, a big energy bill,* and around 11 tonnes of CO_2 for every tonne of hydrogen you make. This is known in the trade as 'grey hydrogen' and, according to the IEA, is responsible for 2.2% of global carbon emissions.

While we've known about hydrogen's energy potential and two-way relationship with electricity for a long time, it was too expensive relative to fossil fuels to get off the ground. It is only now that climate change is forcing us to think about a world without fossil fuels. Renewable electricity will be key to substituting them, but as we have seen, there are limits to what it can do directly. Hydrogen's characteristics make it a great potential partner, but clearly grey hydrogen won't do. Thankfully, we have developed a whole host of clean ways to make it.

---

* To fuel this reaction, you need to syphon off around a quarter of the natural gas you started with.

9

THE HYDROGEN RAINBOW: EXTRACTION METHODS

We have found a multitude of ways to extract hydrogen without any meaningful carbon footprint. Having different colours of hydrogen – green, blue, dark green, pink and turquoise – is a good thing. More sources of supply will mean greater liquidity, supply security and competition.

Unlike fossil fuels, hydrogen can't be dug out of the ground. It has to be forcibly liberated from another molecule. But we are getting better and better at doing that. In fact, we have many ways to extract hydrogen. In an attempt to keep track of all the many possibilities hydrogen has to offer, industry has adopted a loose colour-coding system, sorting these possibilities into broad groups.

There's *grey* hydrogen, mentioned in the previous chapter, extracted from natural gas or coal via steam reforming, which is how we make most of the hydrogen used today. During this process, carbon dioxide is released to the atmosphere.

If you capture and store the carbon dioxide, rather than releasing it into the atmosphere, then you are making *blue* hydrogen.

If you use renewable energy sources to make hydrogen – mainly renewable electricity to split water – then you get *green* hydrogen.*

Dark green hydrogen is produced from biomethane plus carbon capture, which can have negative emissions.

There is also *pink* hydrogen, which is made with electrolysis using nuclear power.

Finally, *turquoise* hydrogen is made from natural gas via pyrolysis, Pyrolysis is heating something up, without burning it, to the point where it splits into its constituent parts. Here the process results in hydrogen and solid carbon as by-products. Although this seems promising, it is still at a very early stage. It is worth noting that the carbon footprint of Blue and Turquoise hydrogen will depend on whether any of the natural gas used to make them is lost along the supply chain.

I fully expect more colours to be pressed into service soon, as new technologies emerge. The techniques and kit for making different types of hydrogen are at different levels of maturity, electrolysers, for green and pink, are not difficult to make. Capturing and storing carbon dioxide is also a well-proven technology, although not without its challenges – not least of acceptability. Pyrolysis – which has great potential because it gives you a way to use methane without needing to store that pesky CO_2, is stuck in the lab. And that goes for the pyrolysis of waste plastics to create hydrogen too.

Different vested interests champion their own favourite colour of hydrogen in a bid to either extend the life of their existing business or add value to their products. So, the

---

* It is also sometimes labelled green if it comes from steam reforming of biomethane, or gasification of solid biomass, but those are not likely to be major sources, and in this book we will use green to mean exclusively electrolysis + renewables.

fossil-fuel industry is quite partial to blue; renewable electricity producers like green; the nuclear industry prefers pink. This competition can sometimes generate more heat than light.

Over the timescale we are interested in – the run up to 2050 – the main ways to make clean hydrogen will likely be green and blue. My money's on green hydrogen to take the largest share of the pie because it is truly renewable and will one day be very cheap and easy to make, but I'm not dogmatic about it. There are places in which blue will have an important role to play.

For instance for Russia, sitting on natural gas supplies that will last centuries and that cost so little, it will probably make more sense to make blue hydrogen by capturing, then storing the carbon released in its manufacture. Blue also makes sense as a stepping stone to green because it can very quickly be produced at scale, which gives you the oomph to convert lots of consumers to hydrogen and build the hydrogen infrastructure. As green ramps up, these customers can then switch seamlessly to an even better type of hydrogen.

Having access to lots of different colours of hydrogen also means they will compete, leading to lower overall prices. And if one source trips up for some reason – say your green hydrogen production suffers an outage – you can swap to a blue hydrogen source. Blue hydrogen (or even grey, in exceptional circumstances) can act in lieu of the petroleum 'strategic reserves'* that countries currently hold, increasing supply security.

As all hydrogen is identical, it will have a key advantage over oil. Every oil reservoir is unique, and produces a chemically

---

* US strategic petroleum reserves store 797 million barrels of oil (equivalent to 1,350 TWh), worth something like $50 billion. If you had to store the same amount of energy in batteries, the capital cost of the batteries would be $400 trillion.

distinctive substance (heavy, light, sweet or sour in different combinations) that needs to be refined differently to produce each end product, such as diesel, gasoline, kerosene and fuel oil. Hydrogen, being just H_2, will be a lot easier to trade – creating liquidity in the market.

The fact that all these hydrogens are chemically indistinguishable poses some problems too. How will you know you are really being sold real green hydrogen and not black-market grey? It costs more to make, after all, and is worth a lot more. That's why it is so important to have proper standards of what really does constitute renewable/low-carbon hydrogen, along with guarantees of origin that can prove that what you are buying is the real deal. Luckily there are many established organisations and companies that provide standards, including the ISO (International Organization for Standardization).

With hydrogen buzzing, we are making great progress on building our toolkit to produce it.

Green

Over the last three years, my team and I have spent a lot of time walking the factory floors of electrolyser and component manufacturers. I was keen to invest in this area because I could see it was going to be big. Over the course of 2020 alone, the market value of the UK group ITM Power rose six-fold, while that of its competitor Nel more than tripled. That's because they are gearing up to supply lots more of their machines – the electrolyser market is projected to grow by 10–15% a year to 2030. I also liked the idea of having a position in the nerve-centre of the hydrogen ecosystem, to better understand how the market was developing and to support the ramp up of this enabling technology. The upshot of our analysis was that in 2020 Snam ended up becoming a shareholder and industrial partner of ITM Power and Italian company De Nora.

My early take is that success in the world of electrolyser manufacturing depends on finding the right balance between three conflicting objectives – cost, performance and durability. The more efficient the electrolyser is at converting electricity to hydrogen, the more precious metals it needs – which makes it expensive to buy. The higher performance also wears the thing down faster. Companies that have performed hundreds of thousands of hours of research and development are good at making this sort of call.

There are two main types of electrolyser being made today: alkaline and PEM electrolysers.

The alkaline electrolyser is an old friend, which has been used to make hydrogen for more than a hundred years. Early in the twentieth century, a little over 400 industrial electrolysers were running in Europe, driven by electricity generated by hydro-power, making hydrogen for fertiliser to assure Europe's wheat supply. They were grown-up versions of Johann Ritter's design, called alkaline electrolysers because of the alkaline electrolyte in them. They were few, but they were big. The capacity of an electrolyser is typically measured in terms of the electricity it consumes to make hydrogen. As early as 1927, electrolysers with a power consumption capacity of more than 1 MW were being turned out by the Norwegian company Norsk Hydro, now Nel. The biggest containerised alkaline electrolysers being built these days consume up to 2.5 MW, and there's no technical reason they could not be made bigger. They can be stacked, one against the other and wired together. There was one absolute monster that operated in Glomfjord, Norway, from 1953 to 1991, that had a staggering capacity of 135 MW.

A much bigger job for alkaline electrolysers was to produce chlorine and caustic soda by splitting a sodium chloride solution. This is now a huge industry – caustic soda by itself is a $30 billion market – feeding the chemicals sector. One of the pioneers of this industry was Oronzio De Nora, who founded his self-titled company (and the one Snam now owns a stake in) in a small lab and turned it into a market leader. De Nora made electrodes for caustic soda plants, and now also makes the core component of alkaline electrolysers, known as a stack. We think of the stack as the processor inside a computer, and of De Nora as the Intel of this world.

Alkaline electrolysers have a lot going for them – including the fact that they are relatively cheap – and getting cheaper still. They are reliable and a good way of making a stable supply of hydrogen if there is a stable supply of electricity. They are, however, bulky and the liquid inside them is often corrosive, and it takes time to ramp them up and down.

There's also a new kid on the block, the proton exchange membrane or PEM electrolyser. It was conceived in the mid-1960s, by General Electric, to produce electricity for NASA's Gemini Space Program, before being brought to Earth in 1987 by the company ABB, which turned out the first high-power commercial models.

The PEM electrolyser puts its electrodes directly on either side of that membrane that Ritter did not bother to fit. This membrane – a specially treated material which looks something like ordinary kitchen plastic wrap – only conducts positively charged ions; it blocks the negatively charged electrons.

While the electrons flow around the membrane through an external circuit, forming an electrical current, protons cross the membrane from the anode to the cathode. There they pick up electrons to become hydrogen. This arrangement does away with the bulky (and often caustic) liquid electrolyser, although the membrane itself has to be kept damp.

PEM electrolysis gets going more quickly than alkaline electrolysis, which makes it ideal for harvesting intermittent energy sources like solar and wind. It's smaller than the equivalent alkaline electrolyser, stackable, easier to maintain, more reliable and simpler to operate. It's efficient, converting around 80% of electrical power into stored hydrogen energy; over time that figure will only increase.

There are of course a few downsides. After about 40,000 hours of continuous operation, that magical membrane will have degraded beyond use. This is a fair innings for any machine part, but this one is particularly costly. Even more pricey is the platinum needed to coat one of the electrodes (and the iridium for the other doesn't come cheap either). Platinum is produced mainly in South Africa, Russia and Canada, and its price seesaws wildly. Endless studies have been conducted in order to find an alternative, so far without success, although the amount of platinum required is dropping fast. PEM electrolysis is a young technology, developing quickly, so watch this space.

Siemens has been operating a 6 MW PEM electrolyser in Mainz since 2015 and ITM Power Ltd has secured a joint venture project with Shell to construct a 10 MW PEM electrolyser in the Rhineland Refinery Complex in Germany. REFHYNE, the largest PEM electrolysis plant in the world, is now operational. It will use renewable electricity to produce about four tonnes of clean hydrogen per day, or about 1,300 tonnes per year.

Looking forward, both alkaline and PEM electrolysers are likely to have their market niches. Most likely, due to their rapid response capabilities, PEM electrolysers will be widely used to manage the intermittent renewable energy sources whereas alkaline electrolysers will be used in more industrial applications where cost and reliability are determining factors.

A third type of electrolyser is the solid oxide variation. This conducts all the usual operations of a PEM electrolyser, only at very high temperatures, around 500–900°C. It has the potential to be the most efficient, but it is an early-stage technology. The leader in solid-oxide electrolyser technology is a German company called Sunfire. Other companies are joining in, including FuelCell Energy, which is building a prototype that will generate 250 kg of hydrogen per day.

Pink

If you make green hydrogen using power from nuclear plants rather than solar or wind power, it is known as pink. The trick here is that with the right kind of electrolyser, you can exploit waste heat from your power plant, in the form of superheated steam. Today, the steam from nuclear plants is sometimes used to make grey hydrogen through steam–methane reformation. It would be far better if we made hydrogen by splitting that steam through electrolysis, perhaps using a solid oxide electrolyser. A 1,000-megawatt nuclear reactor could make more than 200,000 tonnes of pink hydrogen a year.[1] Ten of these reactors would generate around 20% of today's US hydrogen demand.

Pink hydrogen could be a new source of revenue for a

nuclear industry that struggles to get new projects online, or, indeed, keep existing ones running. That would be a good thing as nuclear power has the potential to provide low-carbon electricity at scale. In particular, it could be used as an argument for new small-scale modular reactors, or advanced technology such as thorium reactors that could prove to be lower cost. Given hydrogen's flexibility it could also be a way to transport nuclear energy over long distances, and to match steady supply to fluctuating demand.

Blue

Blue hydrogen is made, like grey, out of natural gas, but the process is cleaned up using carbon capture utilisation and storage (CCUS) which is a label given to a small army of tried and tested technologies. All involve capturing carbon dioxide, either from big point sources such as cement and steel plants, and fossil fuel power plants, or directly from the air. The technology is simple enough – you pass the exhaust gas through a solvent, which soaks up the CO_2; then you heat the solvent and collect the CO_2 as it bubbles out. Once it's collected, you can either bury it somewhere, so it can't get back into the atmosphere (although 'somewhere' becomes a political issue in many countries), or use it to make something useful. It can be a feedstock for the production of synthetic fuels, chemicals and building materials, or make the bubbles in sparkling water or beer, or be added to greenhouses to help crops grow.

The cost of CCUS, which can vary from as little as $15 to as much as $145 per tonne of CO_2 (depending on the purity of the stream of CO_2), largely comes from the capture side of things.[2] Transport and storage each add something in the region of \$10 per tonne of CO_2. The challenge is to scale this whole process up from today's level of megatonnes per year to the gigatonne scale, integrating transport and storage, and drive costs lower. The more existing infrastructure and reservoirs you can use, the lower the cost. CO_2 can also be transported via ship and stored in existing reservoirs.

The Global CCS Institute tracks twenty-six operating facilities around the globe able to capture 40 million tonnes of CO_2 per year, and thirty-seven under construction or development. Many of the projects are in the US where they make money in three ways – through federal incentives such as the tax credits (based on section 45Q of the Internal Revenue Code); through additional local incentives, such as those mandated under California's Low-Carbon Fuel Standard programme, and through the sale of CO_2 to oil companies for a process called enhanced oil recovery (EOR). In EOR, CO_2 is pumped into a mature oil reservoir, increasing the overall pressure and forcing remaining oil towards wells.

There are many uses for CO_2, but using captured CO_2 will only get us so far. We need to get rid of so much of the stuff that there aren't enough potential uses we can put it to. The best idea to scale the process up is to just sequester carbon dioxide in disused oil reservoirs. Around some industrial areas, companies are getting organised to capture CO_2 from local emitters, and then store it in large geological sites. Examples include one in the Humber region (Zero Carbon Humber) and one on the east coast of northern England (Net Zero Teesside), both of which have the North Sea as a final destination for the CO_2.

Over the coming few years, and depending on where you are, blue is likely to be the cheapest form of clean hydrogen, costing only about $2.5 per kilogram – and developing CCUS for blue hydrogen could lead to many other benefits.

In many places, CCUS is the most cost-effective way to curb emissions in iron and steel and chemicals manufacturing, and as far as I know it is the only mature technology to curb the emissions of cement production. In other sectors, CCUS has a hugely important transitional role to play, mopping up emissions from fossil fuels that remain in the energy mix. Climate scientists have been saying for years that we will need this technology to reduce our net emissions. We may well need CCUS to suck CO_2 out of the air, having bust our carbon budget. Still, it has struggled repeatedly to get going. Annual investment has consistently accounted for less than 0.5% of global investment in clean energy and efficiency technologies.

As well as cutting costs, we need to refine the technology. Right now, about 10% of the carbon dioxide passing through CCUS systems still escapes into the atmosphere. Using it alongside steam-methane reforming can cut out about 90% of the carbon emissions from the process, leaving 10% to escape, so blue hydrogen is pretty clean but not zero-emission. Other technologies which could be pressed into service (and which are already used in the chemical industry), such as auto thermal reforming, have even higher capture rates, above 95%.

Another reason CCUS has not taken off is opposition from some climate campaigners. CCUS is often viewed as a fossil-fuel technology that competes with renewable energy for investment. But we need all the carbon reduction tools we can get our hands on. There is also a fear that CCUS will be used simply as a kind of greenwash, allowing polluters to carry on polluting. My own interest in CCUS was sparked by Gabrielle

Walker,* who calls it 'the most unloved, unwanted and vilified climate technology of all'. One of the biggest barriers facing the technology, she told me, is lack of trust, and specifically the difficulty some climate campaigners have in countenancing a role for the oil and gas industry in addressing an issue they are perceived to have helped create.

Things are moving, albeit slowly. Since 2017 there have been dozens of announcements for CCUS facilities, mostly in the US and Europe.[3] But even if these projects go ahead, global CO_2 capture capacity will still only be about 0.3% of emissions. Given how important carbon capture will probably be for limiting climate change, the hope is that it can benefit from an association with hydrogen to make its way back up the agenda.

Turquoise

If you heat up natural gas in the absence of oxygen, the molecules break apart, leaving you atoms of carbon and hydrogen. The carbon is in pure form, just black soot, so this means of making hydrogen produces no carbon dioxide emissions whatsoever. As the process begins with natural gas, it could be appealing for countries such as Russia, Iran, Canada and Qatar, which have gas reserves that will last for centuries.

The technical difficulty with methane cracking is the sheer stability of the methane molecule. You need a lot of energy to break it open: temperatures above 550°C or so will work;

---

* As well as discussing climate change with me at length on that hike up the Norwegian mountain mentioned earlier, Gabrielle has been instrumental in building trust on carbon capture and storage.

temperatures between 800 and 1,200°C are best. Some methods use a plasma torch to reach these temperatures. Although there is some ecological cost to extracting the natural gas, the overall process can be pretty clean if renewable electricity is used to power your plasma furnace. Methane cracking by this method would use only a third of the electricity an electrolyser requires to produce the same amount of hydrogen. Or, once you get started, you could make your own power by burning a little of the hydrogen generated, perhaps 15% of the total yield.

Do all this to everyone's satisfaction, and the result is turquoise hydrogen, so-called because it looks greener than blue hydrogen and because, presumably, someone's media department somewhere was stuck for something to do. And it's not a magic bullet: there's still the question of what to do with all the solid carbon that's been made. Though it is admittedly much easier to store than CO_2.

Soot literally smothered one early attempt to make methane cracking industrially viable. It coated the nickel–iron–cobalt catalyst used by chemists at the petroleum company Universal Oil Products. The only solution they could find was to burn the stuff off – creating carbon dioxide. Now novel furnace designs have begun to address this problem. The carbon black and hydrogen process, developed by the Kværner Oil & Gas company in Norway, manages to separate the soot, which the company can then sell. Soot is useful in the manufacture of rubber and paints. Another approach, developed partly by Nobel Prize-winning particle physicist Carlo Rubbia, involves bubbling methane through a molten metal mixture.[4] The carbon collects on the surface of the metal and can be swept away and used as a very useful, high-performance construction material to replace steel and cement in some applications.

Methane cracking is just one species of pyrolysis (to remind you, the name given to any process that heats something up, without burning it, to the point where it loses its nature and splits into its constituent parts). Another kind involves the thermal decomposition of waste at 540–1,000°C, producing gases that can be used to generate electricity or power transport. A proposed plant in Lancaster, California (just north of Los Angeles), will use plastics and recycled paper as a feedstock to produce hydrogen.[5]

At the time of writing, construction is due to begin on the plant in 2021 and it is expected to be working in the first quarter of 2023, when it will produce as much as 11 tonnes of hydrogen per day, 3.8 thousand tonnes per year. California already has a thirst for hydrogen, and the Lancaster plant's output will go to supply the state's forty-two hydrogen fuelling stations.

The company behind this idea, SGH2 Energy Global, claims that hydrogen produced in this way will be between five and seven times cheaper than green hydrogen produced with renewables and electrolysis. Like any waste-based technology, however, the supply of feedstock is limited by what we throw away – and we should really be throwing away less in the future.

If we can use pyrolysis to get hydrogen out of methane, plastic or paper, then why not do the same with water? Cracking water takes more energy than cracking methane (CH_4) because hydrogen hugs oxygen even more tightly than it does carbon. If you do this with pure heat, it requires a temperature of around 2,800°C. The likeliest source of power is a heliostat: a very large field of mirrors, reflecting sunlight into a reaction chamber that looks like it fell out of *Blade Runner 2049*. Then once you've produced your superheated hydrogen and oxygen, you have to keep them from meeting, because if they do, they will explode. The apparent simplicity of this idea, and the significant technical challenges involved in realising it, are like catnip to young academics. More than 300 water-splitting cycles are described in the literature, each with different sets of operating conditions, engineering challenges, and hydrogen production opportunities. Like many other clever ideas of how to make clean hydrogen, the technology of water pyrolysis is promising, but still in its infancy.

Because there's only one type of hydrogen regardless of how it is made, these many production processes will create connections between different strands of the energy system, different regions, different goods and services. Say it is a very windy week in the North Sea and the price of green hydrogen falls. Customers in Europe would buy less North African or Australian Desert Green, Russian Blue or French Pink, so the price of gas, nuclear energy, fertilisers and even bread would all fall. This is why infrastructure connecting customers to lots of different suppliers is so important. It enables hydrogen to act as the great connector in the energy space, with benefits on liquidity and prices. The rainbow of clean colours shows how lively the hydrogen production sector is. Given the amounts of hydrogen we will need, the more ways we have of producing it the better.

IO

HANDLING HYDROGEN

Hydrogen gas, uncompressed, takes up an enormous amount of space. To get around this problem, hydrogen can be compressed and pushed through a pipe or turned into a liquid and transported in ships. In many cases, hydrogen can flow through existing natural gas infrastructure. This makes it almost as cheap and simple to transport and store as fossil fuels, and a great deal more so than electricity.

In April 2019, something quite special happened. Snam, the company I lead, introduced hydrogen into the Italian gas transmission network. The experiment, the first of its kind in Europe, was conducted in Contursi Terme, in the province of Salerno. We created a blend of natural gas with 5% hydrogen and fed it to two industrial companies in the area, a water-bottling plant and the pasta maker Orogiallo. Later, in December, we increased that mix to 10% hydrogen. The result? Delicious pasta decarbonara.

Everything worked perfectly, even with a 10% hydrogen blend. We tested the effect on pipelines, valves and the equipment at the pasta plant, and no equipment needed to be changed. This experiment made the front page of the *New York Times*, and Orogiallo pasta had a boom in sales.

Contursi opened the door to the possibility of blending hydrogen throughout the gas network, or even carrying pure hydrogen.

Blending is clearly a step in the right direction. It will make the gas mixture a little cleaner and it could be the best way to scale up hydrogen production and kickstart the global hydrogen economy. True, receiving methane and hydrogen combined may not suit some customers, so blending is likely to be piecemeal, with different percentages in different areas. Or we could use membranes to separate the H_2 and CH_4 molecules. These exist in industrial gas separation processes but have not yet been tested at the scale we need for the gas network.

But blending is likely to be just a transitional phase. It won't get us to net zero.[*] For that, we'll need our grid to carry pure hydrogen.

A concern was that hydrogen could infiltrate the carbon steel that pipelines are made of, making them brittle. How rapidly that happens depends on the quality of the steel. The softer the steel, the more disordered its atomic lattice and the less damage an extra hydrogen atom can do. Happily, much of the pipeline grid in Europe is made from softer steel grades, where the process of embrittlement is very slow. They are also very thick, which helps. Snam's engineers have now calculated that at least 70% of the lines in Italy are already ready to carry 100% hydrogen at a pressure equal or slightly lower than the

---

[*] Unless the hydrogen is blended with renewable gas (biomethane), but as with all biofuels the quantities of that are likely to be limited.

one we use for natural gas, and remain safe for fifty years. Indeed, the technical specifications for making hydrogen pipes, of which there are 4,500 km in the world,[1] turn out to be largely identical to those used to make natural gas pipelines in Italy.

Where necessary, it is possible to update the pipework. Obviously, this is expensive, but pipes don't last forever, and old parts of the gas network are getting replaced all the time. This is already happening in some places. The UK distribution network, for instance, is being replaced as we speak, from mainly wrought-iron pipes to polyethylene pipes, which happily can carry pure hydrogen.

A cubic metre of hydrogen contains only a third of the energy of a cubic metre of natural gas, but fortunately this does not mean we need three times as many pipelines. Hydrogen has low viscosity, so its flow speed can be higher than for natural gas. With additional compression, you can bring the maximum energy capacity of a pure hydrogen pipeline to 80% of the energy capacity it would have when carrying natural gas.

The giant ball valves that regulate gas flow should also be able to cope with pure hydrogen; but other elements in the transmission network will need to be updated. Among the most costly items, we will need new compressor stations. Overall, however, the whole process is very manageable. The cost of refurbishing a natural gas pipeline to carry hydrogen will only be about 10–25% of the cost of building a new one.[2] According to the European Hydrogen Backbone, a study run by twenty-three European gas transport companies, by 2040 Europe could have roughly 40,000 km of pipelines, 70% of which would be refurbished. The possibility of using the existing network to carry hydrogen is wonderful news for the energy transition, because it means we have a key piece of the puzzle already in place. It is also important for pipeline operators such as Snam because it means our assets will play a crucial part in the energy system even when natural gas is phased out.

In the vault

So, we've compressed hydrogen, and put it through the natural gas pipeline system. Where are we going to store it?

A slightly flippant answer is: *in the pipes*. The volume of gas held in the pipes themselves, known as 'linepack', provides a lot of useful slack for the gas grid. By allowing this volume to vary slightly, substantial imbalances between supply and demand can be bridged for several hours without affecting supply to customers. But on the scale we're thinking of – if, for example, we're going to make hydrogen in summer to use in the winter – we need to be able to bury vast quantities of hydrogen until we need them.

Storing large quantities of the lightest element underground, at pressure, sounds like the height of folly, but in fact we've been doing it for decades. The petrochemical industry in Texas needs a continuous supply of hydrogen to its refineries and their solution has been to store it in caverns. The Chevron Phillips Clemens Terminal in Texas, for example, has stored hydrogen in a disused salt cavern since the 1980s. Meanwhile, in the UK, there are three salt caverns safely storing hydrogen.

Underground gas storage involves compressing and injecting the gas into a cavity of some sort. Gas is released under pressure when needed. For natural gas, underground storage in disused fields is remarkably cheap, at something like $10/MWh even if you only use it once a year.

For hydrogen, salt caverns are likely to be the best option in some parts of the world. The walls of a salt cavern are tight enough to hold hydrogen gas, even under high pressure. Existing salt caverns, with volumes from a few tens of thousands to more than a million cubic metres, happily operate at pressures of 200 atmospheres and more, giving them the kind of capacity ideal for storing the fluctuating amounts of hydrogen generated by wind and solar power.

Existing salt caverns can only take us so far, however, and new caverns will need to be formed by injecting water through an access well and dissolving the salt. This isn't something to be undertaken lightly: solution mining generates large volumes of brine that we'll need to dispose of in an environmentally friendly way.

This will work in the US, UK and Germany, and other countries where geology is generous with its salt deposits. This would provide cheap hydrogen storage even below €10/MWh (depending on how many times a year it is used). Elsewhere, underground storage will be more complicated. In Italy, for example, there are depleted gas fields that we could use for storage – if we could be sure that the hydrogen wouldn't react chemically with the reservoir structure. That's a big if: hydrogen is very reactive and might react with microorganisms, with chemicals like sulphur, even with minerals in the rock. And, of course, it is likely to pick up residues from leftover hydrocarbons. The same goes for depleted oil fields elsewhere.

Various projects are testing if it is possible to store hydrogen in gas fields as a blend with natural gas. The Austrian gas operator RAG, for example, has a project aiming to demonstrate that depleted fields can tolerate hydrogen up to 10%. RAG also has a project called Underground Sun Conversion. The idea is to put both hydrogen and CO_2 into the same storage site, plus a healthy dollop of bacteria, and see whether they react to form natural gas again (a process called methanation), using the storage site as a natural bio-reactor. At the time of writing, the jury is still out on this project. But the idea of creating a sustainable carbon cycle around geological storage sites is very interesting indeed. Storing hydrogen within the CH_4 molecule (methane) allows you to pack a lot more energy into a given volume and get more bang for your storage buck. It also means you can use existing depleted gas fields, no questions asked, because the molecule would be the same as the one they store today and they have contained successfully for millions of years.

. Once your hydrogen is stored as methane, how will you use it when you take it out of storage? Well, you can just use it as a fuel – provided the original CO_2 that was captured won't produce net carbon emissions. But you would need to keep capturing new CO_2 to continue making new methane. Another idea would be to split the methane molecule again when you take it out of storage, just like in the production of blue hydrogen, creating a closed-loop CO_2 cycle.

Many parts of the world do not have handy salt caverns or depleted gas fields. There are lots of other storage options, such as purpose-built metallic vessels above or below ground. An intriguing option is pipe storage. We already know that pipelines can hold hydrogen gas at pressure, but what's to stop you from laying down a series of cheap, standardised pipelines with sealed ends and using them as storage? A kilometre of pipeline (of the same diameter and pressure that you'd use for natural gas) could hold approximately 12 tonnes of hydrogen.[3] Another clever idea is to line a rock cavern with a thin layer of steel. This has been done in Skallen, Sweden, for use with natural gas. It allows you to ramp up the storage pressure to 200 atmospheres, because the rock formation carries the main structural load, and would be cheaper and easier than a massive tank.

Liquid sunshine

On 10 May 1898, the Scottish chemist James Dewar managed to cool hydrogen enough to turn it from a gas into a liquid. (He also invented the vacuum flask – an idea commercialised by two German glassblowers, Reinhold Burger and Albert Aschenbrenner, giving us the Thermos flask.)

Dewar compressed hydrogen under a pressure of 180 atmospheres, and cooled it using liquid nitrogen down to a temperature of 77 kelvin (-196°C). He then let this pre-chilled hydrogen escape through a valve, cooling it further. You can feel this if you blow air through almost-closed lips; the draft is cooler than your normal breath. In Dewar's experiment, this process cooled hydrogen down to about 20 kelvin (-253°C), which at normal atmospheric pressure is the point where it condenses. He produced about 20 cubic centimetres of liquid hydrogen – about 1% of the hydrogen he started with.

The efficiency has improved since, thank goodness, but handling liquid hydrogen is still technically challenging. Refrigeration is expensive. Tanks must be well insulated, because as the liquid hydrogen absorbs heat it boils away; tank material is also an issue, because metals exposed to extremely low temperatures can become brittle. A similar process is today used to liquefy, transport and store natural gas (LNG). There are hundreds of liquefaction and regasification terminals, LNG tankers criss-cross the ocean, and cryogenic trucks take the supercooled gas to refuelling stations, so such challenges can be overcome. We are not starting from scratch. Some are thinking about converting existing LNG facilities to hydrogen.

A lingering disadvantage is that even liquid hydrogen is still very light, about 70 kg per cubic metre. A litre of liquid hydrogen holds about 2.4 kWh of energy, while a litre of petrol holds close to 9.4 kWh. Turn hydrogen into a liquid, and it *still* has a density problem.

Table 5: Comparing the energy density of fuels

Fuel	Energy density kWh/l
H_2 at atmospheric pressure	0.003
H_2 at 200 bar	0.5
H_2 at 700 bar	1.4
Liquid H_2	2.4
Ammonia	4.3
Petrol	9.4

A denser option still is to chemically combine hydrogen with other elements to make a chemical that is easier to carry and store. These are known as liquid organic hydrogen carriers, or synfuels, depending on the intended use. It can go to form synthetic kerosene, for example, which could help aviation and other sectors move to a low-carbon future. The constraint for synthetic fuels that combine hydrogen with carbon, however, will be the carbon. To be zero emissions, this would have to be sucked out of the air.

One of the most promising hydrogen carriers is ammonia, a compound of hydrogen and nitrogen that we mainly use today to make fertilisers. This book is about how hydrogen can save the world, but the truth is, through ammonia, it already has. Without it, many of us would starve.

Plants need nitrogen to grow. 78% of the atmosphere is nitrogen, but it is in the form of tightly-bound nitrogen molecules, and is useless to plants. It needs to be combined with other elements before they can use it, for instance to synthesise chlorophyll.

The Earth's first nitrates were provided by lightning. It superheats the air, enabling nitrogen to combine with oxygen to form nitrogen dioxide, which falls to the ground in rain or snow: the first fertiliser, if you like. Over time, bacteria evolved to perform the same chemical trick without the super-high temperatures. Clovers, legumes and a few other plants then evolved to host these bacteria, earning themselves their own personal nitrogen supply.

A hundred years ago, chemists realised that there wasn't enough nitrogen finding its way into the fields to sustain our food production. The first stab at solving the problem was a wonderful scheme to generate artificial lightning. In 1902, a pioneering hydroelectric plant at Niagara Falls powered thousands of electric arcs, torturing the air to fix nitrogen the way lightning does. It worked, but the yields were too low, and the process too expensive.

Nitrogen dioxide is not the only chemical that plants can use to get their nitrogen fix. Ammonia is made of one part nitrogen to three parts hydrogen, hence its symbol, NH_3. And Fritz Haber, a professor at the University of Karksruhe, discovered how to make it. Working with the talented lab technicians of the German chemical company BASF, Haber created a machine that passed high-pressure hydrogen and nitrogen over a hot catalyst, creating ammonia. He was awarded the Nobel Prize for Chemistry in 1918 for his work.

While ammonia-based fertilisers keep our food supply going, the same chemical is now a promising fuel to clean up shipping. It works well in internal combustion engines, and even in fuel cells (today these usually run on natural gas, but they are being developed for both ammonia and pure hydrogen). Holding 4.3 kWh per litre, liquid ammonia has nearly half the energy density of diesel (and nearly twice that of liquid hydrogen). That's more than enough to make it a viable engine fuel.

Pressurised and stored as a liquid at room temperature, ammonia is also a potential way to store hydrogen. Converting hydrogen into ammonia only to convert it back again might seem strange, but if the alternative is to supercool hydrogen to its liquid state it may well be worth doing. This makes ammonia a very attractive solution, especially when a lot of energy needs to be stored for a long time.

These three methods of handling hydrogen lead to three different products.

1. compressed gaseous hydrogen
2. chilled liquid hydrogen
3. ammonia, from hydrogen reacted with nitrogen

Each will have its market niche. The following chart shows how much it might cost to get hydrogen to Germany in 2050 from different places and using different methods. Compressing hydrogen is cheaper than liquefying it, so that will be the choice where feasible, for instance where there is a pipeline connection to a sunny location. That would be even cheaper than producing hydrogen in Germany through offshore wind power. For longer distance transport, liquid hydrogen could be better, and it may also be the fuel of choice for some truck manufacturers, as it packs an extra punch compared with compressed gas.

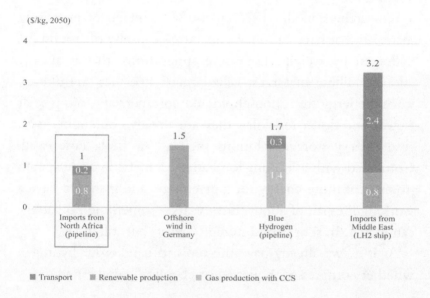

Delivery cost of hydrogen in Germany by 2050

Future generations of cars and trucks could carry hydrogen in an entirely different way, however, absorbed by a solid. The roots of this magic trick reach back to the beginning of the nineteenth century. English chemist William Hyde Wollaston discovered the metal palladium in 1803. Soon it revealed a highly peculiar property: it could absorb large amounts of hydrogen. The discovery was at the time hardly more than a scientific curiosity. Now, it's the basis for a promising new technology.

The ideal storage material, the sort you'd be happy to carry around in your car, for example, would have to be able to store at least 6.5% hydrogen by weight (meaning a 75 kg tank could store up to 5 kg of hydrogen). So chemists are looking to synthesise a material that will hold a large volume of hydrogen but at the same time be able to release it as a gas under mild conditions.

Researchers in the 1970s identified a range of promising materials that have a room-temperature capacity of around 2% hydrogen by weight. For some applications, that is already enough. These materials, called metal hydrides, are already storing energy from household and commercial rooftop solar panels, and dollar for dollar they are already contending with rival battery storage solutions such as the Tesla Powerwall. Compared with a lithium-ion battery, a metal hydride system stores a lot more energy for a given size, and it should have a working lifespan of about thirty years – something no battery can match. In short, solid storage is neat, but still niche.

As it is, we already have the tools to handle our hydrogen whether compressed, liquefied or chemically combined.

II

USING HYDROGEN

We have also developed tools to wield hydrogen's stored energy in a variety of ways, making it a remarkably versatile fuel.

Now you have chosen a clean way to make your hydrogen and are able to store and carry it in a convenient form, how might you unleash its stored energy?

Well, you could just burn it. Hydrogen makes an excellent fuel. You can heat your home or cook a meal by burning hydrogen, the same way many homes burn natural gas. The turbines at the heart of gas-fired power stations need relatively little modification to run on hydrogen instead of natural gas, so burning hydrogen can help keep our power grids from becoming dangerously imbalanced, especially over season-long timescales when we might face weeks with little solar or wind power. The jet engine is a close cousin of the gas turbine, so hydrogen can even fly us between continents on clean airliners. The main waste gas, as with fuel cells, is water vapour.*

---

* Water vapour is a greenhouse gas, but that is not a worry. It does not have a powerful effect when it is low in the atmosphere. And unlike carbon dioxide it doesn't build up in the atmosphere over decades and centuries. It will stay up for about ten days, on average, before coming back down as precipitation.

If you've locked your hydrogen up in a compact chemical form, such as ammonia, go ahead and burn that too, perhaps to drive a container ship. We'll see more of all these applications in the next part of the book.

This is all great if you want heat or motion, or if you have a power station handy, but there's another tool that makes hydrogen even more versatile, by converting its stored energy into electricity. The fuel cell is essentially an electrolyser that works in reverse: instead of using water and electricity to make hydrogen, it uses hydrogen to make electricity and water. It performs essentially the same chemistry as a flame: combining hydrogen with oxygen to make water and release energy; except that in a fuel cell the energy goes into electricity instead of heat.

That means fuel cells can compete with batteries. As long as chemicals flow into a cell, electricity will flow out, more or less indefinitely. This is in sharp contrast to a battery, which has all of its chemicals stored inside. Once it has converted its chemicals to electricity, a battery is dead until recharged or (heaven forbid) you throw it away.

We've already met this technology, in the nascent forms devised by Christian Schönbein and William Grove. But early fuel cells were impractical, not least because of the formidable temperatures and pressures encountered when running the things, which meant that they tended to break down a lot. Francis Bacon – not the philosopher, but a descendant of the same family – built the first practical fuel cell in 1932. Bacon, being an engineer, created an altogether more robust piece of

---

So even if we pump out a lot more water vapour, the atmospheric concentration will increase only by about ten days' worth of emissions.

kit that could cope with high temperature and pressure, the forerunner of today's alkaline fuel cells. His unlovely 'Bacon cell' (a name that, to everyone's secret disappointment, never caught on), full of corrosive potassium hydroxide, was first used to power a welding machine. In that same year, Harry Karl Ihrig, an engineer for the Allis–Chalmers manufacturing company, demonstrated the first fuel-cell vehicle: a 20-horse-power tractor.

Today's fuel cells are better still, largely thanks to the development of new materials. How the electrodes have evolved could practically fill a book in itself. Molten carbonate cells, with magnesium oxide pressed against the electrodes, were a huge step up from anything Bacon had to hand. In their turn, they were soon superseded by very thin Teflon-bonded, carbon-metal hybrid electrodes. Progress continues, now directed mostly at reducing the amount of rare metals required.

Alkaline fuel cells are great for off-grid power, for example where a solar panel feeds an electrolyser to make hydrogen by day, and the fuel cell keeps the electricity flowing after dark. This could be especially important in developing regions with little electricity infrastructure.

For more heavy-duty work we have solid oxide fuel cells, already in demand to generate electricity for factories and towns. They have very high running temperatures (between 700 and 1,000°C) and take eight to sixteen hours to turn on or off, so they only make sense to run for baseload, in an always-on state. The trick with a solid oxide fuel cell is to make the most of its high operating temperature. The steam it produces can be channelled into turbines to generate more electricity. This is why these devices sit at the heart of some of the best combined heat and power (CHP) units currently available. Another technology, molten carbonate fuel cells, also run hot and are highly efficient, and more than 260 MW of these devices are installed around the world for CHP and grid support.

The proton exchange membrane or PEM fuel cell complements alkaline and solid oxide cells. Hydrogen gas under pressure enters on the anode side, and encounters a catalyst, usually a piece of cloth or carbon paper coated with tiny particles of platinum. When a hydrogen molecule comes in contact with the catalyst, it splits into two protons and two electrons. The electrons pass from the anode, through an external circuit where they do useful work like powering your car, and then into the cathode. There they react with the protons and oxygen from the air to form water.

PEM fuel cells are much smaller and lighter than alkaline cells, for a given power output, making them the choice for hydrogen vehicles. Currently, hydrogen cars need about 30 grams of platinum to use as a catalyst in their PEM fuel cells, which is five to ten times as much as traditional vehicles use in their catalytic convertors. This could become a serious problem if the industry scales up, raising demand for the precious metal. Some researchers are trying to make do with less platinum, for example using finer particles to increase the active surface area; other groups are investigating alternative catalysts such as carbon nanofibres with metal oxide nanoparticles.

PEM fuel cells aren't cheap: those precious metals, gas diffusion layers and bipolar plates make up 70% of the system's cost. Nor do they last forever: turning them on and off repeatedly eventually degrades the membrane. There's also a design wrinkle: their membranes must be kept wet, but the cell operates at around 80°C. Consequently, some sort of hydration system is often required to stop the membrane drying out.

Particularly good news for PEM fuel-cell manufacturers is the speed at which the data-centre market is growing. The US now needs nearly 40 GW of backup power capacity for data centres, and that has seen double-digit growth for years, a trend that's likely to continue. That means a doubling of capacity between 2020 and 2025. As big technology companies are trying desperately to decarbonise, hydrogen will be their greenest, cheapest option for data-centre backup. And once they're in place, why only use them for backup? Use them as your main generator, and you can use the waste heat coming off the fuel cells to drive a heat pump, cooling down the server buildings.

An estimated 45% of US data centres could use hydrogen fuel cells as backup power by 2030. By then we're expecting that industry leaders like Amazon, Apple, Facebook, Google, and Microsoft need to add 1,500 MW of stationary power capacity. By 2050, we expect two thirds of data centres to be equipped with fuel cells.

Because fuel cells can also run on methane (natural gas) and generate power more efficiently than a power plant, they can be installed today to electrify many end uses, and await the arrival of hydrogen to the home and the factory. This is a no-brainer for people wanting to upgrade their energy systems without locking in today's technologies. I am planning to install one in my home because methane costs a third of the price of electricity at home, so a machine that allows you to convert from one to the other could save a lot of cash.

There are also solid oxide devices designed to run both ways, as fuel cell or electrolyser. As we will see, these reversible fuel cells are going to be vital for space exploration, helping to power spacecraft and keep astronauts alive on interplanetary journeys, but they will also be useful in homes, to shuffle energy between electricity, heat and hydrogen at will.

And that completes our hydrogen toolkit. Many ways to make it, many ways to use it. Pumps, pipes and ships to move it; tanks and caverns to store it.

Perhaps most importantly, our toolkit lets us convert between hydrogen and electricity, through electrolysis and fuel cells. This means hydrogen can benefit from cheap and plentiful renewable power, and in turn benefit the renewables sector by opening new markets to it. Power-to-gas (P2G) or power-to-liquids (P2L), and its reverse back to power, will be a key feature of our new decarbonised energy systems.

This is how hydrogen can transform today's fragile and frag-mented energy system. With this marvellous molecule at our command, we can extend the reach of renewables, tap the deserts and the ocean winds, bind gas and electricity into a strong and flexible hybrid grid, and decarbonise travel, our homes and industry to finally bring the world to net zero and give ourselves a chance to escape the climate crisis.

Part 3

How Hydrogen Helps

12

HYDROGEN AND ELECTRICITY: A POWER COUPLE

Hydrogen and electricity can come together to create a powerful hybrid energy web, making our energy supply greener, smoother and cheaper. The marriage of molecules and electrons will give us access to the best sites for renewables, by carrying the power of sun and wind over great distances. At the same time, it will calm the jitters of intermittency, and give flexibility and strength to the power grid.

The North Sea holds many fine locations for offshore wind. One is Dogger Bank, 125 km off the east Yorkshire coast in the English Channel. Although it is out of sight of land, the sea here is only 15 to 36 metres deep, which is shallow enough for traditional fixed-foundation wind turbines. And the weather is marvellously bad: it's almost always blowing a gale.

Dogger Bank should soon host the first strands of a workable net zero energy web. Investors plan to build large wind farms here, connecting them to an artificial island using relatively short, affordable cables. One or two huge, high-voltage DC cables would carry the electricity to land (the UK and Netherlands, and possibly later to Belgium, Germany and Denmark).

What's really interesting about this project is what happens to electricity when it reaches the shore. Rather than being poured into Europe's power grids, it will be used to make 800,000 tonnes of hydrogen a year through electrolysis. The hydrogen will then be sent through existing pipelines, mainly to the Ruhr area of Germany, to supply heavy industry such as steel and cement manufacture. What's left over will supply a network of filling stations for hydrogen-powered vehicles.

This plan is pretty revolutionary, but Aqua Ventures goes one step further. As you need to turn the power to hydrogen anyway, why not do away with the DC cable from windfarm to land, and put the electrolysers out to sea? That's the thinking behind a massive 10 GW offshore wind-to-hydrogen hub, based off the little German island of Heligoland, which aims to produce green hydrogen from offshore wind and transport it to land through dedicated pipelines.

While you are putting the electrolysers out to sea, why not integrate them directly with the wind turbine? That's what the Oyster Project wants to do, bringing together wind power companies Ørsted and Siemens Gamesa, electrolyser maker ITM Power and the consultancy Element Energy. They will develop and test a MW-scale compact electrolyser, toughened to survive the marine environment, integrated with a single turbine, with a desalination system to enable it to electrolyse seawater.

These three projects are among the earliest glimmers of cooperation between molecules and electrons. This power couple is coming together to transport and store renewable electricity from one of the cheapest renewable sources in Europe, using existing pipeline infrastructure where possible, limiting congestion on the power grid and providing customers who need hydrogen with the cheap, decarbonised sort.

Catch the weather

One of the challenges with electricity is long-distance transport, which is expensive and wastes energy. If you avoid this by installing renewables close to home, you miss out on the best resources.

Take offshore wind. If you want to produce it close to shore and save on transport costs, you will have lower wind speeds and much less space to play with. If you go further offshore you can tap into a more reliable gale and have enough space to put in truly monster wind farms, but transporting power to shore will be more costly.

Hydrogen solves that dilemma, because putting hydrogen in a pipe carries much more energy much more cheaply than an electrical cable, especially where existing networks can be repurposed. Europe has around 200,000 km of high-pressure gas transmission pipeline. We also have gas pipelines connecting North Africa with southern Europe, and reaching out to oil and gas installations in the North Sea, which can all be easily adapted to carry hydrogen.

This won't be expensive. The European Hydrogen Backbone study estimates that because of all the infrastructure that we already have, transporting hydrogen via pipelines will cost only around €0.10–0.20/kg/1,000 km. That is around one eighth of the cost of carrying energy through the electrical grid. It is also low compared with the production cost of hydrogen, which today is something like €4–5/kg for green.

True, you lose some of the original renewable power when you generate green hydrogen, as electrolysers are something like 70% efficient. But turning renewables to hydrogen can still be cost-effective, especially where you use it in sectors that are technically challenging or expensive to electrify. Hydrogen by pipeline will also be appealing where it's difficult to build additional power lines, or where local populations don't want them.

The ability to carry renewables over long distances opens up some of the best resources in the world: the North Sea, as we've seen, but also the deserts. Back in 1986, and in a post-Chernobyl hunt for sustainable power sources, Gerhard Knies, a German physicist and founder of the Trans-Mediterranean Renewable Energy Cooperation, calculated that in just six hours the world's deserts receive more energy from the sun than humans consume in a year. The Sahara Desert is the world's sunniest area. Its 9.2 million km^2 (more than twice the size of the European Union) enjoys 3,600 hours of sunshine a year.

The Sahara is also one of the windiest areas on the planet. In Morocco, Algeria, Tunisia, Libya and Egypt, some areas have wind speeds comparable to those in the Mediterranean, Baltic and parts of the North Sea.

The potential is staggering, yet plans to export renewable power to Europe have struggled to get off the ground. Witness the changing shape of the Desertec project. Conceived in 2009, it started off as a plan to generate 100 GW of renewable power in North Africa at a cost of €400 billion. It was certainly a large-scale and complex project, and one reason it failed was the high cost of electricity transport. Now in its third incarnation (after a second phase when it focused primarily on domestic demand in North Africa), it has evolved to consider transporting energy as hydrogen.

I think that's a great idea. As we have seen, it is cheaper than transporting electricity. However, there will be some energy lost in the conversion process, and more if you wanted to convert the hydrogen back to electricity. One idea to get around that, in the short term, is to go virtual.

Virtual reality

A plan we at Snam are calling PPWS (put the panels where it's sunny) involves a kind of virtual export of renewable power from North Africa to Europe, using existing pipeline infrastructure. So how would it work? The first part of the project, similarly to the Desertec concept, involves installing panels in North Africa, which is around 80% more productive for a given cost than panels in Germany. It is much sunnier, land is less expensive, and large areas are available for giga-installations, which are much cheaper to lay down and maintain.

Once that cheaper renewable power is produced, however, it wouldn't be physically exported as electricity or even hydrogen. Instead it would displace natural gas now used to generate power locally, often in old and inefficient power plants. The displaced gas could be transported to Europe with no additional infrastructure (the following map shows existing pipeline links in the area), and used to generate power in more efficient European power plants.

Overall, the savings would be staggering. Moving €10 billion of solar energy investments today from central Europe to North Africa would generate 80% more renewable power (18 TWh compared with 10 TWh). At the same time, it would displace 4.3 billion cubic metres of natural gas to Europe, where power stations will squeeze out 7 TWh more electricity than would have been possible in North Africa. This is a great

Map of existing pipelines
© Geo4Map

trade deal that cuts carbon emissions by about 40% compared with the scenario in which the original €10 billion are invested in panels in central Europe. It would also support Europe's political strategy to foster economic development in neighbouring countries, contributing to the growth of their vital energy sectors.

This virtual power swap is an easy win, but only takes us some of the way to net zero – we'd still be burning fossil fuels. The next stage would be to physically export renewable power in the form of hydrogen. For Europe and North Africa this will be relatively simple: as we have seen, existing pipelines under the Mediterranean can be adapted to carry hydrogen up to 100%.

There are still some operational questions. How do you clean panels after a sandstorm? Where do you get the water for electrolysis? Some things may be more difficult than we think, others may be easier. For instance, desalination of seawater for electrolysis only adds $0.01/kg to the finished cost of hydrogen (including disposal of the salt).

It is clear that transporting the desert sun as displaced natural gas, and later as hydrogen, would be a great way to access some of the best renewables in the world. From my own conversations, I know that policymakers in North African countries are beginning to think along these lines. Now we await the actual investment decisions that would move a project like this from spreadsheet-land to real construction sites.

Balancing act

We can use the capacity and flexibility of the gas grid to manage other limitations of renewable electricity, such as intermittency. This isn't such a big problem today, remember, because much of the electricity in Europe is generated by natural gas, which is easy to switch on and off to keep demand and supply in balance.[1] But as we decarbonise, there will be very little power generated by flexible fossil fuels, so we will need new shock absorbers: some other way to steady our power supply and keep the lights on.

As we increase our reliance on intermittent renewables, we will also be using electricity to power more sectors, with different demand profiles. So we risk seeing a growing mismatch between when renewables are available, and when we need them.

If we wanted to solve this issue within the electrical grid, we would need to increase storage, in the form of batteries, pumped hydro and the like; and then we would also need to strengthen the grid, adding many more transmission lines to cope with the sheer volume of power. All these extra cables and batteries would be expensive. They would only be worth installing if they were going to be used often, for instance to smooth over electricity production from daytime to the evening. But what about all those less frequent events? A windless week? The cold snap that covers solar panels in snow?

The molecule grid is on hand to take up the slack.

Already, the gas grid can store some renewable energy, not in chemical form, but mechanically. It acts like a spring, because you can compress the gas to a pressure anywhere between 5 and 70 atmospheres. If your compressors are electric, that flexibility can be transferred to the grid. At times of excess renewable production – say one sunny noon in Southern Italy – you can use the compressors to squeeze more gas into the pipeline, increasing the pressure. When power is scarce, simply switch off the compressors and let the gas expand. You can add another layer of flexibility by making compressor stations dual fuel, able to run on both gas and electricity. Snam is doing this so that we can optimise our power demand to help out the electricity grid.

When the gas grid carries hydrogen, this coupling becomes much more effective. You turn renewables to green hydrogen, send that through the grid and burn it in power plants for peak production, just like natural gas, or convert it to electricity in giant grid-scale fuel cells. True, this loses a lot of electricity along the way – maybe 60% of the power that you started out with. But even so, it may still cost less than having batteries and pumped hydro and lots of expensive new power lines, depending how often you need the backup.

For an idea of how the costs stack up, look at the following chart, which compares the options of smoothing intermittency using batteries, and smoothing it by turning power into hydrogen, storing it, and then turning it back into electricity via a fuel cell. The two bars on the left show you how the options compete when intermittency is daily, while the two on the right show you what happens when you need to store power once a week. The main difference is in the battery cost, which goes from $120/MWh when you use it every day to $700/MWh when you use it once a week. In the daily case, the lower cost of hydrogen transport and storage is more than offset by the cost of turning power into hydrogen and back, but in the weekly case the round trip is clearly worthwhile. This is why you can't just use batteries and pumped hydro to smooth out intermittency in the power grid. You need some dispatchable power for infrequent or seasonal events.

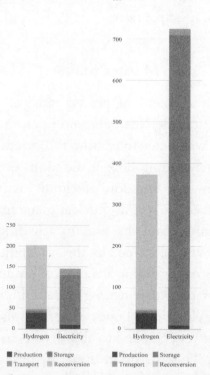

Comparison between H_2 and electricity energy storage.
Left: daily cycle $/MWh. Right: weekly cycle $/MWh

You can also reduce the strain on the electricity grid by making your end customers dual fuel. Homes, offices and factories will one day be equipped with connections to the electricity grid, connections to the hydrogen grid and reversible electrolyser/fuel cells to switch from one to the other. Snam is working with Accenture, Microsoft and Cisco to develop artificial intelligence systems to seamlessly optimise between the two grids. As we have seen in Chapter 5, power prices fluctuate a lot, and when a power station tries to sell you power at £4,000/MWh, a fuel cell that can produce power from hydrogen would be a handy thing to have. All of this will reduce costs as renewable energy will always be delivered through the cheapest pathway.

Market maker

By straddling the world of power and gas, hydrogen also creates connections between commodities, technologies and companies. It will determine how future energy prices are set. On a power-trading desk in London, every second the price of power is a function of clouds, wind, natural gas prices, coal prices, CO_2 costs, nuclear maintenance, transport and storage options and the like. In a hydrogen world, the price of power will be capped by the availability of hydrogen, whether from storage or imports or local production, because you can always turn that hydrogen back into electricity or substitute it for electricity. Power prices will also be supported by hydrogen, always on hand to absorb low-cost electricity to put in storage. This is a good thing, because it will tend to smooth out energy prices.

Hydrogen is also changing the dynamics of the energy sector

– in a good way. As a rule, utilities don't really collaborate, because they stem from regional or national monopolies; they compete with each other for total ownership of an asset. As I found out when I moved from utility Enel to Eni oil company, oil companies do collaborate, but mainly with other oil companies to share the risks involved in building billion-dollar projects.

Now, because of hydrogen, energy companies are no longer working alone. In Italy, Snam is working with Terna, the electricity grid, on joint scenarios modelling the future of energy. The Dutch gas and power grids Gasunie and TenneT have published a study on how hydrogen and renewables could fully decarbonise Holland. Eni and Enel have signed an agreement to bring hydrogen to refineries, as have Ørsted and BP. Total and Air Liquide in France are launching a hydrogen-focused investment platform.

This new interconnected cooperative business model helps simplify today's confusing alphabet soup of energy measures, as everyone rushes to translate their numbers into a system their partners can understand (we are mainly using MWh, as in this book). It also brings the staid energy world a bit closer to the ecosystem approach that Amazon and Apple have used to connect consumers – and our wallets – to lots of different suppliers. These cross-fertilised businesses often give rise to new ideas. An interconnected approach, that combines deep specialisation with a multi-disciplinary capacity to connect the dots, has become an essential skill in modern energy companies. We are working with schools and universities to develop cross-sector energy transition courses and degrees. This broad approach required reminds me a little of what was going on here in Italy 500 years ago. Leonardo da Vinci, at the same time naturalist, inventor and artist – both birdwatcher and wing-builder – embodies the Renaissance. It is my hope that we can bring a little of this multi-disciplinary magic to the hydrogen transition.

13

THE NEW OIL

Global politics has been shaped by oil and gas, while renewable power has mainly been a local business. By opening up the long-distance transport of sunlight and wind, hydrogen will become the new oil, shaping the energy world anew and giving a second chance to those who missed out on the resource lottery the first time around.

Energy and global politics are inescapably intertwined. To pick just one amoung countless examples from history, Winston Churchill nationalised the Anglo–Iranian oil company, the forerunner to BP, to ensure control over the supply of oil to the Royal Navy.

Hardly a year goes by without the European Union making some diplomatic move in an effort to free itself from its reliance on Russian natural gas. (With limited success; in 2010, Russian gas supplied 21% of the EU's needs. It now supplies 34%.) We've had a long-running 'energy cold war', in which the US courted Saudi Arabia and its oil reserves, while Russia cosied up to hydrocarbon giant Iran. And now we have a new narrative emerging, driven by the discovery in the US of huge shale oil and gas resources which have turned it from an energy importer to an energy exporter. As a result, we are seeing a rapprochement between Russia and Saudi Arabia, which are working together to try to rebalance a market flooded with American oil and gas. Meanwhile, China has been building influence through its Belt and Road initiative, through which it finances infrastructure and energy projects in the developing world to gain access to resources, and now Europe and the US are mulling an alternative initiative to finance infrastructure around the world.[1] The inescapable relationship between energy and geopolitics was showcased at US President Joe Biden's Leaders Summit on Climate, which took place in April 2021, and which brought together forty world leaders including long-term allies the UK and Europe, but also Russia, China and Saudi Arabia.

Energy is such an important driver of global politics because without it we can't do very much. We need it to keep the lights on, to grow food, to make things, to get around. And its role could soon become even more profound. With the rise of automation and artificial intelligence, cheap and skilled labour will no longer provide a major competitive edge on the global playing field. Instead, cheap energy will. As labour costs come to mean less, and energy costs mean more, the relationship between energy costs and economic growth will become increasingly close. For a taste of what's to come, just think about those who are mining Bitcoin in Siberia, where natural gas is ultra-cheap. Bitcoin mining is so computing-intensive that it uses serious amounts of energy. Globally, it consumes 121 TWh a year, more than the country of Argentina.[2]

So, it's no wonder that foreign ministers the world over have long been trying to navigate the global order with a map of where the best and biggest oil and gas resources are located, with a particular focus on the Gulf countries, Russia, and the US.

But that is trying to solve yesterday's problem. The politicians should instead be looking at a different map, which shows which countries are going be the most *renewable* resource-rich, with favourable solar and wind potential to make hydrogen with.

Why should a foreign minister trade in the first map for the second? Part of the story is that we are approaching a time when renewables will be cheaper than most fossil fuels at the point of production.[3] But even then, renewable power would mainly be a local business. The real game changer is hydrogen, which allows us to move sunlight and wind around over long distances, through pipelines or on ships. Hydrogen will turn renewable power into the new oil, not only a dominant form of energy but one that can be traded across the globe.

These qualities are sparking the interest of oilmen.

I thought it a sign of the changing times when I got an invitation through the post to a dedicated hydrogen panel at the world's largest energy conference, CERA week (a Houston-based gathering where you can spot the odd Stetson-wearing wildcatter).

Making renewable energy global and tradeable could radically change the balance of economic power. It is a promising prospect for any country with large renewable capacity, including many African nations, the Middle East, South America and Australia. And we don't just need a new map for energy supply: we need one for energy demand, too, as some areas of the world are set to see rapid demographic and economic growth.

As we superimpose the new green oil supply onto new demand centres, we should bear in mind that pipelines are cheaper than shipping liquid hydrogen and ammonia (unless you need liquid hydrogen at the other end). So, the hydrogen market will look a bit like the gas market, with regional systems connected by pipelines, and then some liquids being shipped globally.

Big game in Africa

Africa will be the pivot point of the new hydrogen global order. It will have vast resources. At the risk of repeating myself, the Sahara Desert is the world's sunniest area. Also, on the demand side, Africa will be an attractive and growing market. Over the next eighty years, the continent's population is projected to triple in size to 3 billion, with Nigeria becoming the second most populous nation on Earth, ahead of China.[4] As early as 2030, Africa will count five cities with more than 10 million people, and twelve more over 5 million. We will see vast renewable development to feed this huge and growing energy market, with extra to export.

The effect of this revolution on the continent cannot be overestimated.

For one thing, solar resource distribution is a lot fairer than that of fossil fuels. African countries such as Somalia and Ethiopia, which didn't fare well in the resource lottery the first time around, have a second chance to produce their own energy. That's clearly to be welcomed and will see the balance of power within the continent shift somewhat.

At the same time, electricity lines are relatively expensive to build, and Africa is very large. If you add to that the fact that energy needs in Africa are less seasonal than those in northern Europe, we may see a greater role for distributed energy systems, with homes providing their own energy via solar panels or connected to smaller, local grids. I think there will also be a big role for hydrogen in supplying local energy needs, mainly supporting storage and energy security.

Generating clean energy to power Africa's own development should be the priority. There's a long history of predatory, extractive deals to live down. Renewable development should benefit, rather than harm, the local population – which means being very careful about land use (just because it is empty it doesn't mean it doesn't belong to someone already), water resources and the like. It also means ensuring that African nations benefit from new jobs, training, the development of local supply chains and greater energy security.

Extra for exports

Other sunlit and windy places on our map coincide neatly with current oil and gas producers, which is great news because it offers these regions a new lease of life as we phase out fossil fuel production. It also means we can use some of the same infrastructure again. Oil-producing countries like hydrogen as it lets them exploit their large solar or wind resources, as is already happening in North Africa, the Persian Gulf and Australia.

Saudi Arabia is powering ahead on hydrogen. The world's largest green hydrogen project to date – courtesy of renewables giant ACWA power and industrial gas company Air Products – will be sited at Neom, the 'new model for sustainable living' in the north-west corner of the country. The project is a $5 billion behemoth, integrating 4 GW of renewable power from solar and wind sources (which would cover an area of 80 km^2), turning them into 650 tonnes a day of hydrogen through

ThyssenKrupp–De Nora technology, and then adding nitrogen to make ammonia for export around the world.

Australia, too, has already decided to use its solar power to make ammonia for export. More sunlight per square metre strikes Australia than just about any other place, and powerful winds buffet its southern and western coasts. There's much money to be made in exporting that sunshine to Japan and South Korea, which rely on imported fossil fuels and have limited potential for generating renewable energy of their own. Both countries realise that their best way of reaching net zero is to substitute fossil fuel imports with hydrogen, and Japan is aiming to get 40% of all its energy from hydrogen by 2050. There are already at least three Japanese-backed projects in Australia – Kawasaki Heavy Industries and Australia's Origin Energy are collaborating on a 300 MW electrolyser in Queensland, Iwatani Corporation is working with the Stanwell Corporation in Gladstone, and Mitsubishi Heavy Industries is investing in the Eyre Peninsula Gateway project in South Australia.

Pipeline politics

As we've seen, pipelines are the most efficient way to carry hydrogen over long distances. And pipelines shape geopolitics just as much as resource availability, because they tie producers and consumers together for the very long term. This permanent connection limits both sides' negotiating power somewhat (although you can get around that by having more than one supply source, and more than one export route). But it also limits how badly the countries at either end of the pipeline are prepared to fall out with each other, knitting them together with a subterranean connection even through the most turbulent times.

In a previous incarnation, part of my remit was to run the gas import pipelines from Russia into Europe, and I was always fascinated by the thought that even at the height of the cold war, gas supplies had remained safe. I was also running the Greenstream pipeline from Libya to Sicily during the civil war in 2011, which saw forces loyal to Muammar Gaddafi fighting with troops supported by foreign powers seeking to oust his government. It wasn't the most tranquil period but never did anyone seriously seek to target the gas export route. The pipeline from Egypt to Israel (Peace Pipeline) was operational for many years, bringing Egyptian gas to Israel. And now, following the big gas discoveries in Israel, the molecules in this key strategic asset are flowing the other way.

Given the volumes of gas discovered in the eastern Mediterranean, there's long been talk of a new pipeline project to bring this to Europe. It looks like a tall order, to be fair – 1,900 km away, mainly underwater. But so do lots of the other projects that have or are being built. Turk Stream cost $13.5 billion and is 930 km long, while Nord Stream 2 cost $10.5 billion for a total length of 1224 km.

The real benefit of this project would be as a future hydrogen route. The Middle East – so blessed with hydrocarbons – is poised to win the resource lottery twice, as the Neom hydrogen export project testifies. I have spent time in both Abu Dhabi and Saudi Arabia, where energy executives are committed to diversifying away from oil.

Europe doesn't have much in the way of fossil fuels and is reliant on imported energy. This has caused a lot of soul searching, especially over the relationship with Russia. Eastern

European nations are buying expensive liquefied natural gas from the US, willing to pay a premium to avoid being dependent on Russia. Germany has come under a lot of pressure to halt work on the Nord Stream 2 pipeline, which will double its direct link to Russian gas supplies.

On the back of this history, there is some rather wishful thinking about how the renewable revolution is going to change Europe's relationship with its neighbours, as the continent strives for energy independence. I think pursuing independence is a good idea, up to a point, when it hurts efficiency.

Hydrogen-hungry Germany

Say every country embraced renewable electricity as its national strategy for energy transition. That sounds laudable, but the inefficiencies involved could bring the policy to a grinding halt. Solar panels in the Black Forest produce for the equivalent of around 1,000 hours per year, while sunnier North African countries yield almost double that.

But my main concern with full energy independence is that it creates more problems than it solves. Remember – energy independence does not free us from the need to engage with our neighbours. Those who need energy are dependent, but so are those who sell it. Algeria, Libya, Egypt and the Gulf countries have many new young people with ever-increasing demands. That puts a lot of pressure on government spending, which is largely financed by proceeds from the sale of oil and gas. What will happen to these states as revenues from hydrocarbon production start declining? There is a risk that this would disrupt the fragile balance of the region, drive immigration and undermine security. True, by buying their energy, we gain exposure to potential domestic instability in these countries. But we often underestimate how vital the energy trade is to both producers and consumers.

Meanwhile, if we don't create a global market, we run the risk of local energy costs being vastly different as the efficiency of renewable energy varies from territory to territory. Unless we even out some of those imbalances in an equitable way, we will create a patchwork of states struggling to trade across desperate economic imbalances.

So, I am very much in favour of Germany's pragmatic approach to decarbonisation. It isn't an easy riddle to solve. Germany is Europe's industrial powerhouse and needs cheap energy to make cars. Today, it imports around 60% of its energy needs, and relies heavily on domestic nuclear power, lignite and coal, which are set to be phased out. Nuclear power became politically untenable after the Fukushima nuclear disaster in Japan in 2011, and is set to leave the energy mix by the end of 2022. The Coal Exit law was passed in July 2020 and foresees the closure of all coal-fired generation by 2038 at the latest. At the same time, partly on the back of the Dieselgate scandal of 2015 (when Volkswagen admitted to cheating on US emissions tests), many German cities have banned older diesel cars.

The upshot is that a lot of energy sources are phasing out at roughly the same time, in a country that is already heavily reliant on imports. Clearly, part of the slack will be taken up by domestic renewable production – which has great potential offshore, while onshore it will be constrained by land-use concerns and NIMBYism.* But whichever way we cut it, Germany will need to import a lot of energy, and it will have to be green.

No wonder, then, that Germany was among the first to pinpoint the need for renewable imports, and to put them at the centre of its hydrogen strategy. €2 billion of the €9 billion that Berlin is spending on hydrogen is earmarked for outside Germany. Its officials have visited countries in North and West Africa, the Middle East and even Australia, where they've launched a joint feasibility study on a very long Wasserstoffbrücke (hydrogen bridge).

Among the most striking projects is a plan to build the world's largest hydroelectric power plant in the Democratic Republic of Congo. This could see huge amounts of green hydrogen exported from Africa to Germany. At 44 GW it would have twice the capacity of the Three Gorges Dam in China.[5] It would also submerge land where thousands of people live today, so no surprise that this is a controversial project.

The big guns

If Africa and the Middle East are the new sweet spots of the hydrogen economy, heavyweights China, India and America are not far behind.

Could China be a trailblazer for hydrogen, as it has been for renewables? Installed solar capacity in China went from 4.2 GW

* 'Not in my back yard' – a phenomenon which sees populations broadly supportive of policy goals but unwilling to countenance panels and turbines near where they live.

in 2012 to over 250 GW[6] in 2020 and is forecast to hit 370 GW by 2024 – double the capacity the US is expected to have at that point. As well as harvesting more solar power than anywhere else on the planet, China's manufacturing capabilities, now focused on churning out 60% of the world's solar panel production, could be turned on electrolysers. And there's no shortage of potential hydrogen demand. The government has already announced plans for 100,000 fuel-cell vehicles by 2025, rising to a million by 2030 – a drop in the ocean for vehicles but a giant leap for hydrogen demand. As the government seeks ways to meet its 2060 net zero target, hydrogen is featuring prominently.

India, marketed as the place where everything is possible, is already a land of energy paradoxes. Its hydropower resources – Himalayan glacier meltwater – are already being used to make green hydrogen that's cheaper than coal. Meanwhile, people are still burning agricultural waste to cook, contributing to terrible air quality, pollution and millions of premature deaths. The hope, for India and for many developing countries, is that they will be able to leapfrog from one of the oldest energy systems in the world to one of the newest, rather like mobile phones in Africa leapfrogged fixed-line telecoms (and even retail banking, thanks to the success of money transfer services such as M-Pesa).

The third giant in our triumvirate is the US, which as always is blessed with abundant resources. It has desert sun and ocean wind. It also has the most successful business sector in the world, a climate of corporate invention and a flexible and skilled labour market. Energy innovation has long come from the US – the oil age started with the Spindletop oil gusher in Texas in 1901 – and it has so far been the only place in the

world to make a success of shale oil and gas production, mobilising hundreds of thousands of workers, with an entrepreneur-
ial spirit that no other country has been able to replicate – not
even centrally-managed China.

Clearly, turning spreadsheets into reality is not a straightfor-
ward process. There will be political risks, issues of water availability, snags in the electrolyser manufacturing process, logis-
tics, and sandstorms. But this is a huge opportunity. The ease
of transporting hydrogen means that it can create a fluid global
energy market, linking continents, carrying solar and wind
power from wherever it is most plentiful to where it is most
needed.

As Alan Finkel, Australia's former Chief Scientist has put
it: 'The most marvellous application of hydrogen of all is the
ability for us to continue what we've been doing for hundreds
of years. Ship energy from a continent where it is plentiful to
the continents where it is in short supply.'[7]

14

MAKING MATERIALS GREENER

As the population grows and urbanisation gallops ahead, the world will be more and more hungry for steel and concrete to build with, as well as plastic, fuels and food. Today, these are made through carbon-intensive processes. Renewable power can't always ride to the rescue, but clean hydrogen can step in – as a straight swap for polluting grey hydrogen, and to replace the fossil fuels used to smelt iron and generate intense heat.

The cityscape stretches beyond the horizon. Multi-level road-ways for self-driving vehicles loop out towards a line of tower-ing skyscrapers. The great cranes and docks of a busy harbour are visible behind the domed profile of a concert hall. Vertical farms grow vegetables in automated greenhouses. A drone buzzes overhead, carrying someone's weekly shop across 50 kilometres of suburbia. And this massive city of the future is built on the lightest of foundations.

By the end of 2050, two thirds of the people on the planet will be living in cities. That's 2.5 billion more than in 2021. Housing all these people will mean building four cities the size of Paris every year from now to 2050. That rests on the assumption that we'll have concrete enough, and steel enough, and plastic enough, and food enough and energy enough. All sustainable, and all at the right price. Because although they may look futuristic from the outside, those 2050 buildings and the goods that fill them will be largely made of the same materials that we use today – steel, plastic, concrete, ceramics and glass. That's a staggering amount of raw material, which we will need to make entirely through renewable and low-carbon processes.

Of course we would need most of these materials even if population growth didn't lead to urbanisation. But cities of the future are our trial by fire: if we can't build them sustainably, then we can't live sustainably.

Today, making most raw materials is a hot and dirty business, requiring enormous amounts of fossil fuels both for feedstock and to generate temperatures over 650°C – what industry calls high-grade heat. Industry accounts for roughly 22% of global CO_2 emissions today.

Within industry, roughly half the emissions come from steel, cement and plastics. Other emitting industries include mining and quarrying, construction, making vehicles and textiles.

Given the amounts of materials that we will need to build all our new megacities, along with the roads, bridges, and rail tracks between them, we have little chance of avoiding the full impact of climate change unless we find a new way to make these materials.

That could be tough. Green electricity, so useful in our homes and in our cars, doesn't cut it in manufacturing industry. You can't use it to replace the molecules used in most chemical processes any more than you can replace a banana with a battery. And while it is possible to reach high temperatures with electric heating, it is difficult and costly.

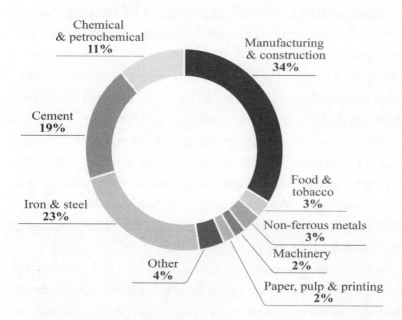

Total estimated CO_2 emissions in industrial sectors, 2019

Green steel

'Whenever I hear an idea for what we can do to keep global warming in check – whether it's over a conference table or over a cheeseburger – I always ask this question: what's your plan for steel?'[1]

Bill Gates

Steel is at the top of the agenda. It is crucial to our lives – in buildings, bridges and cars, and lots more besides. And it is very dirty.

To make steel from iron ore, you first have to extract the iron. Iron atoms are bound up with oxygen in the ore, and separating them requires a reducing agent: a chemical that will grab the oxygen away from the iron. In a blast furnace, iron ore is heated to about 2,000°C and reduced by burning coke (a fuel made from coal). The coke generates carbon monoxide, which acts as the reducing agent, producing a brittle carbon-rich metal known as pig iron. The exhaust from this process, called works-arising gases, or 'WAG', is an unlovely soup, rich in carbon dioxide and toxic carbon monoxide. It finds all kinds of uses on site, and you often find it being piped over to neighbouring concerns, who use it for power generation and methanol production; but the upshot is a lot of CO_2 venting into the atmosphere. Blowing oxygen through molten pig iron takes out some of the carbon and turns it into steel – releasing a bit more CO_2 in the process.

Each year we use more steel: nearly 1,900 million tonnes for the year 2019, which was up by nearly 3.5% from 2018. Every tonne of steel emits about 1.85 tonnes of carbon dioxide on average, so that adds up to 3.5 gigatonnes of CO_2, or about 9% of global emissions.[2] By 2050, as cities rise and expand across the world, demand is projected to grow by up to 40%.

Scrap steel can be recycled in an electric arc furnace, where an electric current melts it. Steel is about the only construction material that we can recycle indefinitely, and it won't lose its mechanical properties. As a consequence, an estimated 85% of stainless steel is recycled once it reaches the end of its life – an excellent example of a circular economy – and a well-developed market has grown up around the scrap trade. The electricity can come from renewable resources, which would make this process very clean.

But recycling is limited by the amount of scrap that's available, so we need to find a new, sustainable, way to make iron and steel from ore. There are some new ideas in the works to do it directly through electrolysis – melting iron ore and passing a current through so that the oxygen bubbles to the surface – but these are still not widespread technologies.

A much more promising approach is direct reduced iron (DRI), sometimes called sponge iron. In this process, ore is heated to between 800°C to 1,200°C, below the melting point of iron, and a reducing gas is pumped though. Today this is usually done using a mixture of hydrogen and carbon monoxide called syngas, made from coal. This is still polluting, but less so than a blast furnace, because the lower temperature means less fuel is needed, and you can more easily reach these temperatures with electricity. In the Middle East, where it's cheap, natural gas is used as the reducing agent. That is somewhat cleaner again – but still doesn't get us where we need to be.

Pure hydrogen does. It is an excellent reducing agent. Oxygen in the iron ore will combine with the hydrogen to produce water instead of CO_2. Pellets of sponge iron can then be heated up in an electric arc furnace to make steel. So there you are: an entirely green production pathway for virgin steel, as long as the electricity powering that electric arc furnace and the hydrogen in the DRI process are generated from renewables.

Germany's Thyssenkrupp and SSAB, a Swedish steelmaking company, are among several pioneers exploring this process. SSAB has built a pilot plant in Sweden which aims to be a step towards fossil-free steelmaking. Meanwhile multinational giant ArcelorMittal Europe is developing its DRI capacity.

We still need to cut costs. At the moment, the green steel from SSAB's pilot plant is likely to be around 30% more expensive than ordinary steel. Steelmakers are particularly ill-equipped to shoulder extra expense because they compete internationally, with wafer-thin margins. Luckily, governments are on hand to help – not least with the initial capital invest-ment which steel companies need, because you can't just use the old furnaces for DRI, you need to build new plants.

It helps that the current generation of blast furnaces in Europe and the US is coming to the end of its useful life. In Europe, this is true for industry more generally: up to half of cement, steel and steam cracker plants in the EU27 will need major reinvestments by 2030.

It also helps that governments are pouring money into the economy as they emerge from the Covid pandemic; steel mills are a great recipient for this cash because turning them green would preserve jobs and give the hydrogen economy some oomph. A significant share of Europe's post-Covid recovery fund is going to its steel industries so they can switch to hydro-gen and electric arc furnace production without going bust.

Even assuming steel mills get an initial push towards green production facilities, they will still need to pay more for their energy, as hydrogen is set to be more costly than fossil fuels for some time to come. How can they hope to compete and make money?

A carbon price would help. That's an additional cost imposed on CO_2 emissions, which makes cleaner alternatives more competitive. Europe already imposes a CO_2 price on some sectors, through the Emissions Trading System (ETS), and prices have been rising. In 2020, they averaged around €30/tonne. By April 2021, the price of carbon dioxide in Europe was edging close to €50/tonne. We may well get €80–€100/tonne by 2030, and higher than that in 2050. So in Europe, given how expensive CO_2 emissions are going to get, hydro-gen-based steel is likely to compete on cost between 2030 and 2040.

If we want companies to start sooner, one scheme that's being considered is the Carbon Contract for Difference, which basically means when a plant switches before ETS makes it competitive to do so, governments would bridge the remaining gap.

Steel competes internationally, of course. So even if a European steel mill decides to pay up and produce more expensive green steel, what's to stop dirty steel from stealing its market share?

That fear has prompted economists (and industry groups affected by the rising ETS price) to argue for a carbon tax at the border to be charged on the CO_2 emitted in the making of the product outside the bloc, which would level the playing field. This proposal, which I wholeheartedly endorse, would pile pressure on international companies to also decarbonise – and save the CO_2 tax when they import to Europe.

Meanwhile, if the steel industry invests in green production processes, it may be able to sell this new clean product at a premium price. The auto industry could provide willing customers, as Volkswagen and Toyota and others aim to eliminate carbon emissions from their entire supply chains. The extra cost for green steel would not be prohibitive in the context of a car. Look at it this way. A €40,000 car is made using a tonne of steel. Today, that would cost $600. A tonne of green steel would cost $780. The extra $180 is only 0.5% of the total cost of the car. The prospect of cornering the automotive market is a sizeable carrot for steelmakers looking to modernise their plant.

Plastic plans

Now think back to our city of the future, and zoom in to one of those skyscrapers, into a single flat and its massive American-style fridge. Like much else inside the building, it is made of plastic. Despite the recent consumer backlash, plastic use is set to grow – driven mainly by higher standards of living in the developing world. Today, each American uses 139 kg of plastic a year. In the Middle East and Africa, it is just 16 kg.[3]

If we do nothing, increased plastic use will create as much as 4 gigatonnes of annual CO_2 emissions by the mid-century.[4] Recycling and reduced use will be vital to curb these emissions. My daughters have a major issue with single-use plastics and especially straws, and they are right. Plastic bottles should go, and so should food containers, coffee cups and those annoying little plastic stirrers that come with them. That could cut projected emissions by half.

Hydrogen will be vital to mop up the rest. To see how, take a look at the three ways in which plastics emit CO_2: extracting and refining the fossil fuels used as the raw material, the high-grade heat used to process them, and end-of-life disintegration or incineration, where the carbon they contain is released.

Hydrogen can provide the high-grade heat required. It could also be used as a feedstock, along with CO_2 captured from the air, to make green plastics that have never seen the inside of an oil drum.* The key to the process lies in designing sophisticated catalysts to speed up the rate of the chemical

---

* You can also use biofuels but, as we have seen, the scale of anything bio will be constrained by raw material availability.

reactions. It could be a couple of decades before we see this reach commercial scale.[5]

And what about the incineration? Well, that can actually produce hydrogen, and even carbon in a usable form. In 2020, a team of researchers from the UK, China, and Saudi Arabia developed a process for converting plastic waste into carbon nanotubes and hydrogen gas.

A concrete block

Our future cities will also clamour for concrete, which is made using another big emitter, cement. Today, cement-making is responsible for 4% of global CO_2 emissions. Unfortunately, getting rid of that is even harder than it is for steel. When you burn limestone to make the calcium oxide you need to make cement, that unavoidably releases CO_2. And unlike steel, cement is difficult to recycle.

There are low-carbon cements being developed in labs, but none at meaningful scale. Alternative building materials include new entries based on the carbon captured through pyrolysis, and old ones, such as wood, given a new lease of life. Wood is a great way to store CO_2 captured from the air. It is also being made more solid and stable through new construction methods – basically sticking slabs of wood together, in various modern takes on plywood. The world's tallest building made this way is in Norway, clocking in at eighteen storeys, and an eighty-storey project is being considered in Chicago.[6]

However, we won't be able to do away with cement any time soon. Hydrogen cannot clean up cement directly, but it could lend an indirect hand. If the blue hydrogen industry can give CCS a kick-start, that could then be used in cement manufacture, potentially providing low-carbon concrete. This seems like the most realistic pathway to cleaning up cement today. The captured carbon may even prove useful in construction materials.

The stuff of life

While our cities eat steel and concrete, humans also have a carbon-generating diet. The FAO estimates that, in order to feed a bigger and wealthier population, we will need to increase food production by 70%.[7] Simply increasing arable land to a similar extent is out of the question, as that would have a disastrous impact on climate and biodiversity. We will need to increase crop yields; and that means more fertilisers.

The most important of these is ammonia which, as we have seen, has already saved the world. Thanks to the Haber–Bosch process, over 100 million tonnes of atmospheric nitrogen a year are fixed into ammonia, used to make fertiliser for farms across the planet.

Today, ammonia is produced using hydrogen as a feedstock, 31 million tonnes of it per year. That hydrogen is grey, made from steam-reforming natural gas. It is relatively cheap – costing some €0.60 per kg to generate, plus the cost of the 4 kg of natural gas you use to make it, putting the overall price at about $2.50–$3 per kg today. But, as we've seen, it generates intense carbon emissions. Green hydrogen, here, is a shoo-in solution. It will be more expensive, at least at first, but it doesn't require any modification of the production process.

So fertiliser companies are likely to be among the first adopters of clean hydrogen. Yara, the world's largest fertiliser producer, is powering ahead with green hydrogen pilot schemes. It has joined up with wind power producer Ørsted to substitute 10% of its grey hydrogen with green at a plant in the Netherlands – a project that would be equivalent to taking 50,000 conventional cars off the road.

Industry isn't usually what we think of when we think about climate change. Most of our attention is focused on power generation and transport. But industry is vitally important because it is big, because it is hard to decarbonise, and because it may well be the area that gets the ball rolling for hydrogen. That's especially true where hydrogen is already used as a feed-stock, as in fertilisers and refineries: a surprisingly big market, already amounting to more than 70 million tonnes of hydro-gen a year, worth something like $130 billion. Simply using green hydrogen instead of grey is a good way to start the revolution.

15

SOLVING SEASONALITY

The huge energy demands of winter heating require seasonal storage, which electricity cannot provide, and would overwhelm today's electrical grid capacity. But they can be met by a certain green gas.

You might think that life in Italy has a constant backdrop of golden sunlight and al fresco *aperitivi*. Not so. Winters in Milan are so cold that I wear a ski jacket over my suit to go to work, and cup my hands around my takeaway matcha latte (yes Starbucks has finally opened in Milan). I have to keep my home heating on throughout the winter months, and even then I throw on an extra jumper.

In much of the northern hemisphere, winter heating consumes a huge amount of energy. And in Europe, at the moment, that energy is largely provided by natural gas, which heats 42% of all homes. In total, Europe's gas network provides an estimated 90 million homes with natural gas. In Italy and the UK, which both have well-developed networks and cold winters, gas heats a much higher share of homes: 70% in Italy and as much as 85% in the UK.

Overall, in both countries the gas grid provides about three times the energy of the electricity grid over the course of a year. Winter peak demand on the gas grid is around five times higher than peak demand on the electricity grid. And a very cold snap can make the gulf wider. The 'Beast from the East', which hit Europe in 2018, added 12 TWh to Italian gas demand over two weeks, equal to an extra full day's demand.

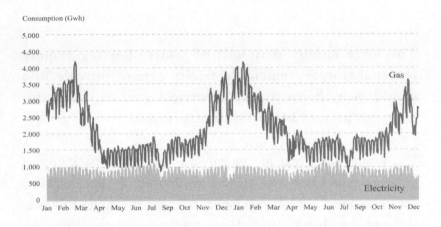

Gas infrastructure is essential to cover seasonal peaks

The heft of winter heating means it accounts for a lot of the CO_2 we emit – about 20% in the UK overall. And, of course, the share is much higher if we just look at cold cities. Around 42% of all the CO_2 emitted by New York City, for example, comes from keeping homes warm.[1] So we must clean it up. But how? The huge seasonal swing means that it is difficult to imagine decarbonising heat simply by switching to electricity.

Our current electricity grids and storage facilities are simply not strong enough to cope with the demand arising from a wholesale switch, even allowing for the fact that electric heating, where you install it, is more efficient than gas. Nor were they designed to be: the gas network is shaped the way it is precisely to cope with seasonal energy peaks. In many areas the capacity of the gas grid can be up to ten times that of the electricity grid. Replacing it with an ultra-high-capacity electricity grid carries a hefty price tag.

Just as problematic is how our winter demand for heat can be met directly with renewable power. Europe's generating capacity from solar power falls through the floor come winter. And even if we made enough electrical energy in summer to tide us over a hard winter, how would we store it? As we have seen, the quantity of energy that we would need to store during the summer to offset the additional demand needed in the winter is five times what we produce and consume during the summer months. Using batteries would be unfeasible and ludicrously expensive, because for seasonal heating we would be using each battery only once a year (if we discharged it on a winter's day, we would need to wait until the following summer to charge it). And pumped hydro won't fill the gap either, as there aren't enough mountain lakes and reservoirs to spare for that purpose.

In Europe, we will want to electrify some heating. For example, when buying a brand new house, getting a heat pump installed may not be a bad idea. These new-fangled sounding devices are actually old and extremely reliable technology. Essentially, they're inside-out fridges, moving heat from the colder outside to the warm inside of a house. They're little miracles of efficiency, and a good way to decarbonise heating in some areas, especially where it isn't so cold. It probably also makes sense to switch to electricity when deeply refurbishing a house – which something like 1% of us do every year.[2]

The thing is, though, you're probably not buying a brand new house. You're more likely to be living in an old property. Installing a heat pump in an old property is both impractical and a financial nightmare, requiring heavy insulation and potentially new heat distribution systems, the total cost of which runs to around \$200–\$300 a square metre.* Few people can shell out that sort of money up front, even if the returns are worth it in the long run. In Europe at least, this is a serious problem. Here, roughly 90% of household emissions come from buildings older than twenty-five years, about three quarters of all buildings in the EU. Worldwide, most of the housing stock is old. It isn't easy to ask consumers to shoulder the burden of decarbonising homes – which is why government programmes to incentivise energy efficienct refurbishments are so important. Italy offers a 110% tax benefit, meaning that for every €10,000 you spend on your refurbishment you get €11,000 tax relief over five years.

Electrifying all these old homes isn't just a financial nightmare. It is also logistically impractical. The UK's National Grid has estimated that if the country wanted to start a mass electrification programme in 2025 and end it in 2050, it would need to convert 20,000 homes a week.[3] The best hope, here, is that heat pumps are developed which do not require such deep renovation of the building stock.

---

* Electric heat pumps produce heat at a much lower temperature than gas appliances, so they can't easily be used with radiators. You need an underfloor heating system that allows the heat to be exchanged over a much greater surface area.

Where electrification doesn't add up, what can we do?

In some areas, we may decide it makes no sense for every property to have its own heating system. Why not use the waste heat from electricity generation to heat buildings directly? This is the principle behind combined heat and power. A large electricity generating station can redirect its waste heat directly into hundreds of homes, a system known as district heating. Or on a neighbourhood scale, a combined-heat-and-power unit hardly bigger than an electricity transformer station can generate electricity, and redirect the heat generated into a few surrounding streets. Existing CHP units are already capable of processing hydrogen, as well as natural gas. But that's not a solution for everyone – for starters because the heat doesn't get to you at very high temperatures. And the solution to slightly tepid radiators is expensive insulation, which takes us right back to square one.

Another potential solution is to use green alternatives to natural gas. Why not, for instance, carry on using methane (the primary constituent of natural gas), but get it from renewable sources? This biomethane is indistinguishable from the gas we already use, so it works with existing pipes and appliances. It releases carbon into the air when it's burned, but as this carbon was captured from the air by plants, emissions are very low.

Biogas, made by digesting organic matter using anaerobic bacteria, has been around for well over a century,[4] and anaerobic digesters have produced methane from waste biomass for many decades. We could all be cooking and basking in its warmth tomorrow if only the supply was there. Supply, however, is the besetting problem with this mature and efficient green technology. Biofuels are becoming less land- and resource-hungry than they were, but they still have to be grown and harvested, so we're always going to have trouble finding enough organic feedstock to make biofuels plentiful and therefore affordable, especially for such an energy hungry sector as heating.

Biofuels will recur in our story because they will play an important role in the green economy, especially in the production of green aviation fuel. They're not a bulk replacement for natural gas, though.

One interesting idea is to make the best use of this scarce resource by combining it with electricity in a hybrid system to heat our homes. We would install reversible heat pumps and some insulation which would be enough to provide summer cooling and moderate winter heating, and would be fine for most of the year. And for the really cold snaps a small biomethane boiler would kick in.

Now, of course, we come back to hydrogen, which can be a bulk solution to winter heating. You can produce hydrogen in the summer and store it underground, and then use it in boilers or generators or district heating networks in winter.

As hydrogen is such a powerful and potentially green source of heat, why don't we use it in our homes already? In fact, we did, and not so long ago. The coal gas that powered most of the UK into the 1970s was roughly half hydrogen, half methane, with up to 10% toxic carbon monoxide. It was made by heating up coal in the absence of oxygen. At 400°C–450°C, this gives off a gas that burns with a bright flame. Scottish engineer William Murdoch was the first to make use of coal gas, lighting his house with it in 1792. The idea took off, and Murdoch's employers Boulton and Watt, who made steam engines, started to build small gas works for factories. In 1812, the German entrepreneur Frederick Winsor obtained a Royal Charter to build the world's first public gas works. London's streets lit up, and soon so did streets around England. Within fifteen years, almost every large town in Britain, Europe and North America had a gas works. Coal gas (also known as town gas), heated and lit many countries for decades. It wasn't completely phased out in the UK until 1981, when the last gasworks closed – and Hawaii, Singapore and Hong Kong still use it today.

A blend of natural gas and hydrogen would be a great starting point to decarbonise heat. As I've already mentioned, at Snam, we've worked on a 10% blending trial for industrial use. Elsewhere, similar projects have been trialled for the home. In the UK, more than 650 homes and commercial buildings will trial a 20% hydrogen blend for around ten months, as part of the HyDeploy project. An earlier iteration of this, involving Keele University's private gas network, demonstrated that hydrogen can be safely blended into the natural gas distribution system at concentrations of up to 20% hydrogen by volume, without requiring changes to the network components or downstream appliances. A project in Amerland in the Netherlands safely and successfully injected 20% hydrogen into their natural gas grid for use by domestic consumers, while the GRHYD project in France is supplying a 20% hydrogen mix to around a hundred domestic customers and a hospital.

As we'll see later, this level of mixing could be enough to get the economic ball rolling – but if we want to reach net zero, we eventually have to leave natural gas out altogether and burn pure hydrogen.

This is a highly flammable gas, of course, but it should be possible to use hydrogen safely in homes (we will come to this in more detail later).

Hydrogen boilers already exist, and their cost is projected to fall from about $1,600 to $900 per household by 2030 – similar to natural gas boilers. Installing new boilers in millions of homes would be a major undertaking, but bear in mind that gas boilers are replaced on average every fifteen years anyway. As costs are not so different, there should be an incentive or a mandate for new appliances to be hydrogen-ready now, with boilers that can switch from methane or a methane hydrogen blend to pure hydrogen after a short visit from a heating engineer.[5] Fuel cells for the home will be able to take methane or hydrogen and turn them into heat and electricity.

Before any of this is possible, we have to get hydrogen into the home. We've already seen how existing infrastructure, with a little modification, can carry hydrogen around the gas grid, but that was focusing on transmission lines: the big pipes that connect major sources and users of gas.

We haven't yet looked at distribution pipes – the local network that takes gas from the big transmission pipes and sends it out to individual domestic and commercial users. And here the situation appears to have vast regional variations. The UK is very well positioned because of the Iron Mains Replacement Programme, launched in 2002, which has been upgrading the majority of distribution pipes to polyethylene, which are suitable for transporting 100% hydrogen.

Inevitably, there will be additional costs to meet, not least to replace equipment, detect leaks and monitor hydrogen quality. None of these problems should be hard to address, but it's important to acknowledge that we have no experience in supplying pure hydrogen to domestic users or commercial users outside heavy industry. At present, most of the studies evaluating hydrogen systems are still paper exercises, and we need a big programme of preparatory work before we embrace hydrogen at 100% purity.

At least the paper exercises are promising. In the UK, the H21 Leeds City Gate Project[6] studied how to convert Leeds into a city fuelled entirely by hydrogen. With 1.25% of the UK's population, Leeds is a manageable size, while still big enough to show what's required to develop a hydrogen network. Leeds is also near existing infrastructure at Teesside, where there are geological sites already used for hydrogen storage.

The study shows that switching the network to 100% hydrogen would involve minimal disruption for domestic and commercial customers and require no large-scale modifications to property.[7]

The costs for H21, according the report, would be in the region of £2 billion for infrastructure and appliance conversion, and £130 million a year for operation. Who pays for it is then the big question – as it is of course for the energy transition as a whole. Equivalent studies are now under development in Australia and Ireland, and interest is being shown in China, Japan, Hong Kong, New Zealand and across Europe.

When it comes to practical plans, Scottish gas operator SGN is leading the way, aiming to deliver pure H_2 to 300 homes in Levenmouth. As a first step, it will build a demonstration facility which will allow customers to 'see and experience hydrogen appliances in a home-like setting', so people can get comfortable with the idea before opting in.

This is all essential work. Heating is a large source of carbon emissions today, and a tough one to clean up. Electric heat pumps will do their bit, especially in newbuild or refurbished homes. Hybrid solutions, with heat pumps plus biomethane boilers, could help too. But straight hydrogen will probably be the solution of choice in cold areas, and in homes near industrial clusters which will have hydrogen piped through their distribution systems anyway. How wonderful it will be, one cold January day in 2030, to crank up the boiler and bask in the heat of June's Sahara sun.

16

GREEN LANES

Transport is a major source of CO$_2$ and deadly air pollution. Batteries will work well for some vehicles, but for others they will be too heavy. On land, highly compressed hydrogen offers long range and fast refuelling. It could be the most effective fuel for trucks, buses and taxi fleets, and it may well compete with batteries for passenger cars, too.

I love to travel. I love the speed and the comfort and the freedom. I enjoy driving, and I am a bit of a geek about my car. And I cannot imagine life without flying. Low-cost air travel was the transport revolution of my twenties, when taking cheap flights to new destinations turned me into a true European. Pre-pandemic, I worked across Europe and the US, and emerged from Asian trips energised and with new ideas. I don't think business travel will ever be the same again but I would like to preserve what was truly useful about it. It isn't just me though: many of us love to travel. And many of us need to travel.

Today, transport is intimately linked to oil. Around 95% of energy for transport still comes from oil, and of the 99 million barrels of oil that the world burns each day, between 60

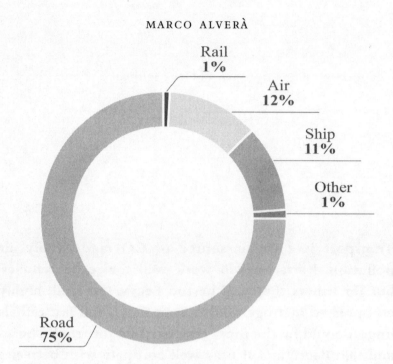

Rail
1%

Air
12%

Ship
11%

Other
1%

Road
75%

Total estimated GHG emissions in transport sector, 2019

and 70% goes into road, rail, ships and planes. Transport's share of global carbon emissions was 21% in 2019 – the vast majority from road transport.

Our energy needs for transport are set to grow hand in hand with population growth and economic development, especially in non-OECD countries. Pre-Covid forecasts saw a doubling of road and aviation kilometres by 2050. Covid-19 may change some of our transport habits, and dent the growth in energy required, but we are still likely to see a lot of extra demand for mobility.

It's all the more important to clean up transport because it generates other pollutants. Fine particles, methane, nitrous oxide, hydrofluorocarbons, perfluorocarbons, sulphur hexa-fluoride and nitrogen trifluoride: all make urban environments unhealthy and lop years off human life expectancy, as well as harming local ecosystems and the global climate. We can choose to travel less, of course, but it's a sad solution in the long run. We can do much better by taking the CO_2 emissions, the pollution and the guilt out of travel. What's the best way to do it?

The most obvious solution to many is biofuel, at least in the short to medium term, because it doesn't require any changes to engines or infrastructure. However, unless you want to restrict the amount of land available to grow food, or start cutting down forests, then most biofuels are limited by the amount of organic waste we give over to their making. Farming seaweed or other algae may sidestep those limitations, but converting algae to biofuel is complex, and the technology immature. So the most realistic candidates for decarbonising transport are, by and large, electricity (plus a battery) and hydrogen (plus a fuel cell). What these two options have in common is that we would be switching virtually all transport to an electric motor.

The modern internal combustion engine is a thing of beauty. It has to be. It's trying to squeeze out all the performance it can out of a ridiculously complex design concept. Lots of tiny fuel explosions generate hot, high-pressure gas that pushes against a piston, turning a crankshaft, which via the gearbox finally drives the wheels. A lot of the energy in the fuel is lost as heat, and there's also a lot of wear and tear on the motor. Only about 20% of the potential energy stored in gasoline is ever delivered to the wheels. And while the roar of a petrol- or diesel-fuelled engine is music to many car enthusiasts' ears, that sound represents wasted energy.

Compare and contrast with the clean, quiet and efficient electric motor, where currents flow through coiled wires, generating magnetic fields whose force turns the driveshaft. There is less friction and no messy explosions, little energy is lost as heat, and the whole contraption needs hardly any maintenance. About 95% of the electrical energy is turned into motion. The electric motor beats the internal combustion engine hands down, so the first lesson of transport decarbonisation is that, wherever possible, vehicles should be driven by electric motors.

The key question is how to store the electricity to power them. Are we going to use renewable electricity and batteries (in a battery electric vehicle, or BEV), or clean hydrogen and fuel cells (in a fuel-cell electric vehicle, or FCEV)?

Well, that depends. Decarbonising transport means working out how to pack a lot of clean energy into a small space, without adding too much weight to the vehicle. And it means doing so at the lowest possible cost, with luck, even below the cost of the oil product (gasoline, diesel, jet fuel . . .) that we are trying to displace. The ideal solution doesn't exist – you will need to make trade-offs between efficiency and range, charging times and the impact on energy infrastructure. And different sorts of vehicles, doing different things, will want to make a different choice.

Electrical engineers, who spend a lot of time thinking about this, will say the battery beats the fuel cell. That's because batteries are very good at doing one specific thing: charging and discharging without losing lots of energy on the way.

Imagine we start off with 100 kilowatt hours of energy produced from solar panels and wind turbines somewhere sunny or windy. We lose between 5 and 10% of this energy as we pass the electricity through the grid, a further 10% from charging and discharging the lithium-ion battery, and finally another 5% from using the electricity to make the vehicle move. Multiply all these losses, and we're down to around 80 kilowatt hours, or 80% overall efficiency. So efficient are batteries that it might be tempting to set aside the idea of hydrogen and fuel cells, which at first glance seem to be wasting energy.

Let's take our 100 kWh of electrical energy produced by a renewable source again. First, we have to turn it into hydrogen, presumably by electrolysis. This is around 70% efficient, which means we've already lost around 30% of the energy we started with. The hydrogen then has to be transported to a refuelling station. The energy used depends to some extent on the transport method, but roughly speaking these processes are around 90% efficient.

Once inside the vehicle, the hydrogen needs converting into electricity. Fuel cells are only around 60% efficient (sometimes more, but let's not quibble). Finally, take out the 5% of energy which is always lost by the electric motor. Put that lot together, and you'll discover that only 36% of the original electricity hits the road; that's 36 kWh out of the 100 we started with – nearly twice the efficiency of the internal combustion engine, but only half that of the battery-powered vehicle. On the face of it, that's a pretty decisive win for batteries. Elon Musk took this perfectly correct argument for a walk at the 2015 Automotive News World Congress in Detroit – but his conclusion, that hydrogen cars 'make no sense', turns out to be wrong.

Free range

Given the higher efficiency of battery electric vehicles, why does the hydrogen plus fuel-cell combo even get a look in? Because we need to look at the bigger picture. The mathematics doesn't stop once your vehicle is in motion. Remember, your fuel source, be it battery or fuel cell plus hydrogen tank, is coming along with you for the ride. And it's while we're in motion that things get interesting. Vehicle batteries are still very heavy for the energy they store: the lithium-ion battery pack in a Tesla weighs over half a tonne. Contrast this with the hydrogen-and-fuel-cell solution. As we've seen, hydrogen is the most energetic element in the universe, holding almost 40 kWh in every kilogram – the highest energy content per kilogram of any chemical fuel, almost three times as much as petrol and over a hundred times as much as batteries.

This means the amount of energy that can be stored is limited not by the weight of the hydrogen but by the volume it takes up in the tank. A fuel-cell vehicle also has to carry around a fuel cell, of course, but these are a lot lighter than their equivalent in batteries. The Toyota Mirai's fuel cell is 56 kg, compared with the hundreds of kilograms of batteries needed for a family car to reach a range of even 200 km.

This all matters because it costs energy to carry weight. A heavier vehicle has to fight against higher rolling friction in the tyres and more friction in the transmission. This gets worse with every traffic light stop and start, and every hill you need to go up. If you don't need to drive very far, you don't need a lot of batteries. A typical battery-electric car has a range of 250–300 kilometres, although some models may double that. If you drive up and down hills, that range will diminish considerably. This 250–300 km range is perfect for city driving.

If you decide to extend the range of your vehicle by piling on more batteries, you soon start to hit diminishing returns. Each battery you add makes your vehicle heavier, increasing friction. That means it takes more energy to move, draining those batteries faster. Each battery you add means the car has to work harder to move; each extra battery takes you less far than the last.

For long-distance driving, especially with heavier vehiclers or in mountainous areas, you might be better off looking at something hydrogen-fuelled. And about time, too. Hydrogen has long been the automotive fuel of the future without ever quite getting in gear.

The very first internal combustion engine, built by Isaac de Rivaz in 1807, ran on it. We had some hydrogen buses on the roads in Germany as early as the 1920s, thanks to the work of German engineer Rudolf Erren, who developed a system that could operate on either hydrogen or petrol at the flick of a switch from inside the cab. The first recorded hydrogen fuel-cell car was the 1966 General Motors Electrovan, which must rank as one of the more terrifying vehicles ever conceived. It was powered by supercooled liquid hydrogen and liquid oxygen, held in two separate tanks with 170 metres of piping throughout the rear of the vehicle, which turned a six-seater into a two-seater. Though it had a range of 190 km, the engineers wisely kept it corralled on company property.

Thankfully, FCEVs have come a long way since then. So while their energy efficiency is lower than that of BEVs, it is worth considering them for vehicles that need range. The Toyota Mirai can hold 5 kg of hydrogen in its tanks, giving it a range of about 500 km, which is comparable to many petrol cars. That might suit the American driving public, as in the US cars travel an average of 12,000 miles per year, far more than in other countries.[1]

Hydrogen's range is one of the reasons why Wan Gang, Chinese minister of Science and Technology, widely credited with having started the EV revolution in China, now says that hydrogen vehicles are the future. China is home to nearly half the world's electric passenger vehicles. In the Made-in-China 2025 industrial roadmap, the government identified hydrogen as a key technology for its electric vehicle market. China wants 50,000 fuel-cell vehicles on the road by 2025, and 1 million by 2030. That's going to require a lot of refuelling stations: China plans one per 1,000 vehicles.

In South Korea about 6,000 fuel-cell cars were on the road in 2020. The country has set a target of 850,000 by 2030. Calling hydrogen power, the 'future bread and butter' of Asia's No. 4 economy, President Moon Jae-in has declared himself an ambassador for the technology and is overseeing $1.8 billion in central government spending to subsidise car sales and build fuel stations. Government subsidies of the Hyundai Nexo model have cut the price in half to about 35 million won ($30,000), and sales of the model surged in response. Hyundai and its suppliers plan to invest $6.5 billion by 2030 on hydrogen R&D and facilities.

The United States is a sleeping giant in this market. In 2003, US president George W. Bush called hydrogen fuel cells 'one of the most encouraging, innovative technologies of our era',[2] but so far California is the only state in the Union that's held fast to that idea. The state has spent more than $300 million in the past ten years funding rebates for those buying or leasing hydrogen cars, building fuel stations, buying buses and subsidising development of hydrogen trucks. All but a handful of the 7,800 hydrogen-powered cars in the US are in California. In 2020, as governor, Gavin Newsom issued an executive order banning the sale of new gasoline-powered vehicles in the state by 2035, the California Energy Commission announced $115 million funding for up to 111 new hydrogen fuelling stations by 2027, which puts the state on a path to achieve economies of scale so the industry will no longer depend on government incentives.

Charging ahead

Another criterion is performance at the fuelling station. Chemically transforming the innards of a battery to charge it up takes a lot longer than pouring some liquid into a can. Waiting for your vehicle battery to recharge can take several hours, which in Italy means a lot of espressos at the motorway café. High-power rapid chargers, such as the Tesla Supercharger, bring that down to roughly an hour, but try and force the pace any more and the battery will degrade.

Long refuelling times don't matter if you're simply plugging in your car overnight, so it's ready for the next day's city driving. And do you really mind waiting an hour for it to recharge? What about getting it done while you're at the gym or in the supermarket parking lot when you're doing your weekly shop?

You might not mind waiting, but the refuelling station does mind. The refuelling station minds a lot. Hydrogen refuelling takes one tenth to one fifteenth of the time fast charging requires, so one hydrogen-refuelling point can serve ten to fifteen times as many vehicles as one fast-charger. That means the hydrogen-refuelling infrastructure requires about ten to fifteen times less space to fuel the same number of vehicles. Admittedly, the upfront cost of building hydrogen refuelling stations is higher, but we can expect these costs to fall sharply.

Slow refuelling times may make it difficult for car fleets to switch to BEVs. Taxi businesses operate in areas where restrictions on emissions make fossil fuel cars undesirable. Battery vehicles would be an ideal replacement, one would think, given they're generally making short journeys. But imagine a fleet of a hundred battery-powered taxis, all parked in one place, all needing a recharge. You'd need a hundred charging points, and a long correspondence with your energy supplier. Sure enough, fuel-cell taxi fleets across Europe are increasing in number and size. Paris, London, Brussels and Hamburg have them, and a fleet in Paris already has more than a hundred vehicles in operation.

Charging time is also one of the main reasons why the forklift truck forms the unlikely advance guard of the hydrogen vehicle revolution. Hydrogen forklifts are as clean as the battery versions (important when they're working in enclosed spaces), they perform well at low temperatures and, crucially, they can refuel in just three minutes. This is important for a vehicle which, although small, has a heavy workload and gets through a lot of energy. While recharging they are out of commission. And for a fleet of a hundred forklifts, a battery recharging station is a big expensive thing. Tens of thousands of fuel cell-powered forklifts are being produced every year, making this workhorse, as of 2020, the largest driver of hydrogen vehicle fuel demand.

New business models are being explored to get around slow charging of batteries. One idea is a technology that lets you

swap your depleted battery for a fully charged one at the refuelling station. This may work well for motorcycles, where
batteries are small enough to carry like wheeled trolley bags.
But while we wait for such ingenious new solutions to emerge,[3]
it seems likely that we'll be taking our battery electric vehicles
for short hops and plugging them into our garage wall sockets
at night.

Gridlock

What happens at the charging station affects the individual car
owner. But it also affects the system as a whole. At the moment
there are relatively few EVs, recharging mostly from wall
sockets fed by the main power grid, but consider what it will
mean to the power grid when many more of our passenger
cars are electric, with more charging points and more electricity consumption. If battery vehicles are going to supplant the
internal combustion engine across the board, then we may
either have to come up with a different way of distributing
electrical energy or treat our power grids to a massive upgrade.

Expanding the electricity grid is expensive but perfectly
possible, so long as we can keep it stable, but to do that we
need to balance the electrical supply between quiet and peak
periods. Fast chargers work against that, by adding peak
demand. In peak times, when people drive to work, return
from work or go on holiday, fast charging will push up grid
load. The Tesla Supercharger has a capacity of 250 kW, equivalent to switching on more than eighty electric kettles at once.
The only way the grid can deal with this is to build additional
peak generation capacity (and transport and distribution). This
is not the end of the world, but it's extra work we could well

do without while we're trying to navigate a major energy transition.[4] There are those who pin their hopes on vehicle-to-grid technology (which we met in Chapter 5), where smart charging of the batteries in EVs stabilises, rather than congests, the grid, but that's still being explored.

Hydrogen service stations are far kinder to the grid. They can produce hydrogen on demand from grid power, or from nearby renewable electricity production; they can receive hydrogen through pipelines, or in compressed or liquid form from trucks. They'll have busy periods and slack periods like everyone else, but they'll handle these fluctuations without making fluctuating demands on the electricity infrastructure. You can also use hydrogen from the grid in a stationary fuel cell to charge electric vehicles, which may be worth doing if connecting the refuelling station to the power grid is difficult.

Putting all of this together, BEVs have by far the highest efficiency and will likely take a big share of the market – especially for urban passenger cars. Hydrogen will make sense where electrification isn't an option, or where its advantages in range, charging time and impact on infrastructure outweigh its lower efficiency.

Light at the end of the tunnel

An early adopter of hydrogen technology is likely to be the grand-daddy of all heavy vehicles, the train.

Trains are the ultimate in heavy land transport. Electrification of the railway system began early; the first example in Great Britain was Volk's Electric Railway in Brighton, a pleasure railway which opened in 1883 and is still functioning to this

day. And after the Second World War, electrification transformed a once sooty and polluting transport network into
everyone's idea of a sustainable form of transport. Or much of
it, at any rate.

Not every train line can be electrified for a reasonable cost.
The cost of resizing tunnels and bridges to accommodate the
new power lines, the cost of electrical substations and the cost
of reinforcing the local power grid, all have to be spread over
the number of trains running on the line. Even on busy main
lines, it can take forty or fifty years to earn back such a large
investment. Lines that don't run so many trains would simply
not attract the investment in the first place.* Seventy per cent
of all rail journeys in the world are still diesel. In the US, only
1% of the tracks are electrified. This is a problem, as diesel
trains are both dirty and expensive, and need to be phased out
for us to reach net zero. So the pressure is on to replace diesel
trains with trains that run on a different, much cleaner fuel.
Trains are big and heavy and go a long way. What better choice,
then, than to power them with hydrogen? They have the room
to store the stuff, and they need every drop of the energy it
packs.

The first 'hydrail' locomotive was demonstrated in Val-d'Or,
Quebec, in 2002. The world's first fuel-cell passenger train,
Alstom's Coradia iLint, entered commercial service in 2018, on
a 100-km regional line in Germany. In an interview with the
news agency Agence France-Presse, Stefan Schrank, a project
manager at locomotive builders Alstom, was bullish about the
prospects. 'Sure, buying a hydrogen train is somewhat more

---

* Switzerland runs an entirely electrified rail network, but it is not exactly a
poverty-stricken country.

expensive than a diesel train, but it is cheaper to run.'[5] In 2022, a fleet of fourteen iLints will be replacing diesel engines in Lower Saxony, travelling up to 1,000 km on a single tank of hydrogen, and reaching a respectable 140 kph.

The UK is not far behind: a fuel-cell-powered prototype called Hydroflex is now running on mainline rail tracks. China, too, is testing light rail and tram systems which use hydrogen-generated electricity for their motive power. As are France, the Netherlands, Japan and California. In Italy, Snam has an agreement with Alstom to develop hydrogen trains.

Trains return to depots which are often located in industrial areas, so they may be able to obtain blue hydrogen from nearby factories, at least until they can find an affordable supply of green hydrogen.

Keep on trucking

Vehicles that need long range, including many buses and trucks, will also be hungry for hydrogen. Hydrogen buses can go more than 500 km on a full tank, while battery-driven buses manage about 200 km on a single charge. Many city buses are electric today – there are 450 in London alone. But longer ranges are attractive even for buses running short routes, as you want to be able to keep going for a long time between charges.

Today a hydrogen bus costs twice as much as a diesel bus. But if the government subsidised the first 3,000 vehicles, according to Jo Bamford, Chairman of the UK's Wrightbus, that would enable factories to scale up and churn out buses that cost the same as their diesel counterparts.[6]

Fuel-cell buses are already running in fourteen European cities, including Aberdeen, Antwerp, Cologne, Oslo and Riga. The

European 'H$_2$ Bus Europe' funding programme will support 600 new fuel-cell buses over the next five years. Toyota and Hyundai already sell fuel-cell components to bus makers. In China, manufacturers including state-owned SAIC Motor, the nation's biggest automaker, and Geely Auto Group (which owns the Volvo Cars and Lotus brands) have been developing buses of their own.[7]

Trucking routes tend to be long distance, so all the arguments above apply: your business will lose a lot of time and money if you have to keep pulling off the road, hooking up to a heavy-duty 350-kilowatt charger and waiting potentially hours for your huge bank of batteries to fill up. The batteries also cut into the load you can carry. For Tesla's semi, the option with 300-mile range (roughly 500 km) needs a battery capacity of 570 kWh which will weigh roughly 4.5 tons. The 500-mile variant requires 950 kWh capacity, or nearly 8 tons. And while battery prices have been plummeting, that's still likely to be quite expensive.

The new generation of hydrogen trucks is nearly here. Daimler Trucks, the world's largest commercial truck-maker, says they will have a full line of hydrogen commercial vehicles ready by 2027. They have recently partnered with Volvo to develop hydrogen systems for the post-internal combustion era. General Motors has teamed up with Navistar to launch a new semi-truck,[8] set to reach the market in 2024. Anheuser-Busch, the world's largest brewer, has ordered 800 fuel-cell powered trucks, burnishing its green credentials. It will be a few decades until our power grid becomes 100% green, and until then, using renewable hydrogen is the only way to physically ensure 100% green trucking.

It's not just about big five-axle rigs. Smaller vans could also favour hydrogen. Renault & Plug Power have joined forces with the aim of becoming the leader in hydrogen light commercial vehicles. And Stellantis (born from the merger of Fiat Chrysler and Peugeot) is planning to sell three models of hydrogen fuel-cell van by the end of 2021.

Not that this is a winner-takes-it-all competition. Eighty per cent of freight in the United States is transported less than 250 miles. At that range, whether you plump for batteries or fuel cells is mostly a matter of taste.

So there we have it. Hydrogen will play a role in rail travel that cannot be electrified, in long-range trucking and for buses, taxis and some passenger cars – with trains and trucks switching first. The demand from heavy goods vehicles is likely to give the whole market a push, because it will help get some filling stations on the road. Trucking routes and trucking volumes are relatively stable, which takes a lot of the risk out of running a hydrogen refuelling station. Even better, trucks will usually return to one or a handful of sites sooner or later – so there's no difficulty in spotting where you need to build those first fuel stations. And there are enough really big companies out there who have the capital to invest in fuel-cell vehicles early on, and get hydrogen transport rolling.

17

GREEN SEAS

Maritime transport uses particularly dirty fuel. It accounts for around 3% of total greenhouse gas emissions, and generates a lot of air pollution. If no action is taken, its emissions are projected to grow by up to 250% by 2050. The noxious exhaust from cargo ships could be cleaned up by replacing their traditional sludgy fuel with ammonia, made using hydrogen.

Long-range ships have to cover thousands of miles before they can refuel, so they need to pack in a lot of energy. At the moment, most big ships burn a noxious porridge known as bunker C, or residual fuel oil. This is the sludge left over after gasoline and other fuels have been distilled from crude oil. This stuff doesn't just generate carbon dioxide but also smog-forming nitrogen oxides, lung-clogging particulates and climate-damaging soot. Go visit a city with a port terminal and you can see the effects clinging to and eating into the very surface of its buildings. Yet 90,000 ships use this fuel, carrying everything from grain to medicines to waste paper and old shoes all over the globe.

No wonder, then, that by 2010, ships were fast becoming the biggest source of air pollution in the European Union. As I write this, emissions of sulphur dioxide and nitrogen oxides from ships are expected to exceed land-based emissions in Europe.

But there is a ray of hope. The industry is regulated by its own international governing body, the International Maritime Organization (IMO), and that body fully recognises the need to decarbonise.* The IMO's target is to reduce global annual greenhouse gas emissions in maritime transport by at least 50% by 2050, with a view to achieving zero emissions as early as possible during this century. It will be an uphill battle given how much demand for shipping is set to grow.

On the other hand, shipping routes are regular, and they are everywhere. Even the least-developed countries, lacking traditional energy infrastructure, have well-established ports. Many ports could build the infrastructure needed to make clean hydrogen, using electricity generated from nearby wind farms or solar farms on land. Many have energy providers next door that are increasingly keen to find uses for all their shiny equipment before the energy transition leaves it stranded. And the areas around ports are usually home to industries that have located there because it is easy to get raw materials, or simply because they are co-locating with other industries.

This means that ports are a great place to find hydrogen demand. Indeed, they are leading the charge on early-stage hydrogen projects. The Port of Antwerp, which is one of the busiest in Europe and hosts potential hydrogen consumers such as chemicals companies BASF and Ineos, and oil major Exxon

---

* Inadvertently, this process has already begun. The IMO has required fuel suppliers to strip the sulphur out of their maritime fuels – a process that commonly uses hydrogen; this has already increased demand for hydrogen and so helped reduce the cost of electrolysis.

has come up with a plan to import hydrogen. The Port of Rotterdam is also looking to import hydrogen, as well as producing its own green (onsite using electricity from offshore wind farms) and blue (putting the CO_2 in the depleted Porthos field in the North Sea), and connecting to a national hydrogen pipeline network.

Better yet, the shipping companies' clients are getting increasingly nervous about the cleanliness of their activities. Maersk, one of the largest shipping companies in the world, has committed to reducing net carbon emissions to zero by 2050. But if its ships are not going to run on bunker C anymore, what should they burn?

It is feasible, at least on paper, to power even the biggest ships with fuel cells powered by hydrogen. In the spring of 2020, the Swiss/Swedish manufacturing giant ABB teamed up with Hydrogène de France (HDF) to develop the sort of megawatt-generating hydrogen fuel-cell systems that big container ships will need if they are going to go fully electric. But how will ships store all their hydrogen, especially on long voyages?

Compressing hydrogen in tanks, the way we do in our cars, is an option, but it will still take up quite a lot of space. Gaseous hydrogen at 700 atmospheres takes up 70% more room than liquid hydrogen. This sounds like a small price to pay for being able to carry the stuff around at room temperature. Unfortunately, it helps to keep pressurised containers small if you want to make them at reasonable cost. A ship carrying many small, pressurised containers runs out of space fast.

If you store hydrogen as a supercooled liquid instead, you have to use about 30% of the energy held by the hydrogen to liquefy it in the first place, but after that you don't need active refrigeration if you are burning the stuff – just decent insulation – because you can burn the hydrogen gas that comes off as the liquid slowly boils. One liquid-hydrogen Norwegian cruise ship is expected to be in operation by 2023. Carrying liquid hydrogen on board is technically challenging and expensive though, and I expect this to remain a high-end option, for luxury passenger ships, for some time to come.

Smaller boats, such as private yachts and the *vaporetti* in Venice, will also run on hydrogen. The sooner the better, at least as far as Venice is concerned. Its ships make it one of the most polluted cities in the Italy despite not having any cars. Of course, vessels that are even smaller, such as water taxis and private boats, may well run on batteries.

A potentially more practical route is to combine hydrogen with nitrogen from the air to make ammonia. With relatively few tweaks, existing ship engines can burn ammonia, emitting water and nitrogen. No carbon dioxide is released, and relatively little in the way of particles and other noxious fumes. Consequently, it's likely to be the most cost-effective way to get green power into long-haul shipping.

Ammonia is toxic, and designers trying to build road vehicles around this fuel have been defeated by safety concerns. At sea, though, the expertise and equipment already exist to handle the chemical, as it is routinely used on board ships to scrub nitrogen oxides out of engine exhaust. Ships that use it as fuel will still need a new set of safety regulations, which could slow down their development.

Alternatively, hydrogen could be combined with carbon dioxide to make a synthetic natural gas or synthetic liquid hydrocarbon. Ships that burn any of these fuels will be able to operate at a small fraction of their current environmental impact. However, converting shipping to synthetic fuel will probably require an extra nudge, a subsidy to make the economics stand up. That's because the fuel would be more expensive than traditional fossil fuel sludge, once you add up solar or wind power production, electrolysis and the cost of the process of converting hydrogen to ammonia or other synfuels.

There is one way shipping can lower emissions right now. When cruise ships are docked at ports, they need a lot of power for their fridges and air conditioning and onboard appliances for thousands of passengers. The port's grid is usually unable to provide enough electricity, so the cruise ships run their heavy-oil fired onboard generators instead, emitting CO_2 and clouds of horrible pollution. A fuel cell at the port could offer cruise ships electricity made from methane, and in future hydrogen, saving much of the CO_2 and all the pollution.

Meanwhile, it is just possible that we could see the return of the hydrogen airship, that pioneer of early twentieth-century flight. The idea is to use the jet stream to propel vast cargo-carrying airships across the world's oceans. The jet streams – fast flowing, narrow, meandering air currents – flow in the mid-latitudes predominantly in a west–east direction at altitudes of 10–20 km, at speeds that commonly reach 150 km/h and can stretch to well over 300 km/h. By hitching a ride on these atmospheric currents, airships could carry the same amount of cargo with a lower fuel requirement and shorter travel time than conventional shipping. They'd be filled with hydrogen, because helium is far too rare and expensive for such an operation; and it weighs twice as much as hydrogen, so it wouldn't be quite as effective.

Even if this daring idea does eventually take flight, passenger services are very unlikely to follow. For that, we must turn to hydrogen's energy once more.

18

GREEN SKIES

Flying may be our great climate crime today, but it could soon be washed clean with liquid hydrogen. That will require reinventing the aeroplane. Hydrogen could be used in a fuel cell, combined with other elements to create synthetic jet fuel, or even carried in pure liquid form to feed hydrogen-burning engines.

The aviation sector accounts for around 12% of the emissions from the transport sector, so just under 3% of CO_2 emissions worldwide. And that percentage is set to grow. Even though air travel tanked in the pandemic year of 2020, it is likely to rebound fast. Passenger kilometres are forecast to grow 4–5% per year over the coming decades. Worldwide, the aviation sector already emits more than 900 million tons of CO_2 every year and, even with efficiency gains from improving technology, that looks set to at least double by 2050. As we bring most of our other sources of emission under control, air travel looks like it could be a stubborn problem.

The simple CO_2 numbers don't even reveal the full climate impact of aviation. As well as CO_2, jet engines emit nitrogen oxides, which create ozone, a greenhouse gas that has an especially powerful warming impact when it is high in the atmosphere. They also emit aerosols and enough soot to blacken glaciers; and they leave behind ice contrails, which, though they sound innocent enough, trap heat a little like a greenhouse gas. The overall climate effect of flying is complex and still being researched, but it is estimated to be around twice what you'd expect from the CO_2 emissions alone.

Airlines and manufacturers have gone to great lengths to clean up their industry. Billions of dollars have been invested to modernise aircraft, with more efficient aerodynamics and engines using lighter materials. Fuel burn per passenger kilometre has dropped by half since 1990. In 2009, the sector set itself ambitious targets that include carbon-neutral growth from 2020 onwards and halving its net emissions from 2005 levels by 2050. But the longer this struggle to clean up air travel continues, the more incremental are the gains. We're nearing a point at which we can't make planes much more efficient.

What's the solution? Certainly, we could reduce the number of flights we take. The Swedes have coined the word *flygskam* (flight-shame) for a movement to encourage people to fly less often, and another, *tagskyrt* (train-brag) to convey pride in taking trains instead. While this simpler life narrative may resonate with some people in wealthier nations, it takes no account of those who seek a better standard of living, including the easy travel so long enjoyed in the West.

Every little helps of course, and we should all do our part, but following the Swedish example and abandoning short-haul flying for the train (or short-hope electric aeroplanes) makes depressingly little difference to the problem. Commuter and regional flight distances account for less than 5% of aviation's

CO_2 emissions. We need change on an altogether grander scale.

We need to reinvent the aeroplane.

Imagine being able to fly anywhere in the world without leaving a trail of destruction. No soot. No carbon dioxide. That's possible with hydrogen.

It's not a perfect solution, but it is pretty good. Flying on pure hydrogen, you would still leave contrails of ice behind you, but their impact on the climate would be much lower. That's partly because the ice crystals from a hydrogen plane are going to be larger and more transparent than those from a traditional, kerosene-powered jet; this means they let more of Earth's infrared radiation escape to space. There would also be a little nitrogen dioxide. At the high temperatures reached in a jet engine, nitrogen and oxygen from the air combine to produce nitrogen monoxide, which later combines with oxygen in the atmosphere to form nitrogen dioxide. Catalytic converters and engine designs can ameliorate the problem, so hydrogen planes should produce around a fifth of the nitrogen dioxide emitted by planes burning kerosene.

Overall, this will be a low-impact form of flying, one that most environmentalists could get behind. This is the dream we're chasing – and with advances piling up fast, I believe it's closer to realisation now than when I started writing this book.

There are two main ways hydrogen can power an aircraft. Burn it as jet fuel,* or feed it to a fuel cell, which then powers

---

* The Heinkel HeS 1 experimental gaseous-hydrogen-fuelled centrifugal jet engine was tested in September 1937.

an electric motor spinning a propeller or fan drive. Both of these options work cleanly and efficiently; each works best with a different size and range of plane.

Fuel cells could well be the answer for shorter flights. German company H2Fly and Singapore-based HES have both designed futuristic-looking four-seaters. One has twin fuselages either side of a central engine pod, the other a string of small propellers along each wing.

Jeff Bezos and Bill Gates, the founders of Amazon and Microsoft, are among the investors in ZeroAvia, a British–American company based in California and Cranfield, England, which has raised $21.4 million from climate funds. ZeroAvia has modified an existing aircraft, a six-seater Piper Malibu, to run on hydrogen fuel cells. In September 2020, the plane took a twenty-minute fuel cell-powered flight from its R&D facility at Cranfield Airport in the UK. ZeroAvia envisages a fleet of ten- to twenty-seat planes flitting between all the small airports that are largely unused today.

Longer flights will need to burn hydrogen in engines. Carrying it around in a gaseous state is probably impractical; even compressed to the 700 atmospheres used in car tanks, you only get 42 grams per litre. If we want to fly on hydrogen, we're going to have to carry it in liquid form somehow.

We could combine green hydrogen with carbon (captured from the air or an industrial plant) to form synthetic kerosene, the main ingredient of jet fuel. This is not a bad option, as it diverts us from using fossil fuels – although planes running on the stuff will still spew out a lot of nitrogen dioxide and other pollutants. And to make synthetic kerosene using hydrogen and CO_2 captured from industry, you'll end up with about half the energy you could have generated with the hydrogen. Use

CO_2 harvested from the air itself and you'll end up with only a third.

Such synfuels could be a useful stopgap measure, along with other forms of sustainable aviation fuel. Some airlines have invested in demonstration plants that make jet fuel out of municipal household waste. British Airways, Shell and Altalto are developing the UK's first commercial waste-to-jet-fuel plant,[1] converting half a million tonnes of non-recyclable coffee cups, food packaging, and even nappies, into clean-burning jet fuel each year. In some places, sawdust is being fermented into sustainable kerosene. But such projects to exploit waste are likely to remain small scale, too small to decarbonise the whole industry.

How about biofuels?

In 2016, the International Civil Aviation Organization adopted a scheme to rein in emissions, aiming for carbon-neutral growth from 2020. In the Carbon Offsetting and Reduction Scheme for International Aviation, sustainable fuels and carbon offset schemes compete, so sustainable fuels will only be widely adopted if they are both affordable and truly effective. The problem is, they tend to be either one or the other. Biofuels based on food crops such as soy and palm oil have an indirect climate impact because they increase food prices and encourage deforestation for new plantations. Advanced biofuels, based on non-food crops such as switch-grass, don't have that effect, but they are expensive. No wonder, then, that sustainable aviation fuel (SAF) is currently less than 1% of total consumed jet fuel. Airlines don't yet have a compelling reason to pay through the nose for SAF.

More challenging, but a lot more exciting, would be to fly on pure liquid hydrogen.

But can we seriously hope to be able to design craft that can carry liquid hydrogen around on long-haul flights at temperatures close to absolute zero? Liquid hydrogen must be kept below minus 253°C to stop it boiling off and venting from the tank. And there's a second major challenge: liquid hydrogen is still astonishingly light, roughly 70 grams per litre. Four litres of hydrogen delivers as much energy as one litre of kerosene. Planes powered by liquid hydrogen needn't be cartoonishly big, but they will need to bulk out to some degree.

Could such planes get off the ground? Certainly. They already have. In the autumn of 1955, a laboratory at US air base Wright Field decided to try flying a plane on liquid hydrogen. They installed a supplementary fuel system in a B-57 twin-engine bomber, and modified one engine so it could use hydrogen or kerosene or a mixture of the two. On 13 February 1957 the plane took off, and climbed using kerosene. Then it switched to hydrogen in the modified engine for twenty minutes. The experiment went off without a hitch.

In the 1980s, Russian aircraft designer Andrey Tupolev (whose planes held seventy-eight world records) decided to perform a similar experiment on a passenger plane. His company rebuilt one of its Tu-154 airliners so that one of its three jet engines could run on cryogenically liquefied gas. The fuselage was extended and a cryogenic fuel tank was built in a carefully isolated and well-ventilated space behind the passenger cabin. The plane, redesignated Tu-155, first flew on 15 April 1988, and proved that hydrogen can produce enough thrust to power a commercial aircraft. Later tests with the same aircraft replaced the hydrogen (an impossibly expensive fuel at the time) with liquid natural gas. The

Tu-155 took around a hundred flights, until the fall of the Soviet Union brought the project to a close.

Further serious research into hydrogen as an aviation fuel had to wait until the early 2000s, when the European Union brought together Airbus and thirty-four other partner companies to imagine how civil aviation could continue to grow until every human being on Earth could fly as often and as far as they wanted without hurting the environment. Confronted with such an extreme challenge, the consortium quickly settled on an extreme solution: power the whole of aviation with renewable hydrogen. The exercise, led by Airbus Germany and dubbed Cryoplane,[2] then became about how to get there.

Cryoplane's design work revealed the potentials and limitations of hydrogen-powered jet planes. A plane that can carry so much liquid hydrogen in super-isolated tanks is going to be bigger than an ordinary jetliner, which means an increase in drag and consequently energy consumption rising by around 10%. However, this is offset by the weight reduction. Even though those tanks and associated cryogenic equipment will add weight, hydrogen itself is so light that a fully fuelled H_2 plane would be around 10% lighter than a fully fuelled regular plane. A compelling picture emerges: the two planes have similar efficiency, but one pollutes the planet a lot more than the other.

There are several possible directions for liquid hydrogen aircraft design. Conventional designs can be adapted to cope with the larger tanks, which look like the best option in the short term. Longer term, the most promising unconventional design is the twin boom system, in which the wings and tail are joined together using the hydrogen tanks. This design, also known as the joined-wing or Prandtl plane, would make a virtue of the desire to separate the tanks from the passenger cabin (for safety reasons, and to prevent heat from the cabin raising the temperature of the liquid hydrogen).

Prandtl planes have structural benefits, as the tail and wings support each other. The design creates a stiffer structure, meaning weight can be reduced. Some researchers claim that it also offers aerodynamic benefits, providing more lift; but the large external tanks also increase drag.

Blended wing-body designs are another option. The lightest shape for a high-pressure vessel is a sphere, which carries forces evenly. However, that would have a larger diameter than a rectangular fuel tank, making it difficult to place the tanks in the area above the passenger cabin of a conventional aircraft – the layout most favoured by Airbus. Blended wing-body aircraft, in contrast, effectively have one much deeper wing, making it easier to install the lighter spherical tanks.

Cryoplane also looked at infrastructure; after all, there has to be some means of making and transporting the hydrogen and getting it loaded into the planes. When Cryoplane's final report was published in 2003, its enlightening conclusion was that, while liquid-hydrogen storage presents some unique problems, they aren't any more difficult or expensive to solve than the problems of handling liquid hydrocarbons. Indeed, hydrogen has one major and massive advantage over jet fuel; leaking hydrogen dissipates quickly into the atmosphere with no harm done. When hydrocarbons leak, they contaminate the soil and the water.

In 2020, the Cryoplane work was back in the news again, as Airbus applied its decades-old knowledge of hydrogen-powered aviation to go for a €15 billion prize. On 9 June 2020, in the midst of the Covid-19 crisis, the French government announced a support package for its embattled aerospace sector. Along with a €500 million investment fund for smaller companies, the package included that startling financial incentive for whoever could unveil a carbon-neutral plane by 2035 and demonstrate it in flight by 2028. Barely three months later, in September, Airbus announced plans for the world's first zero-emission commercial aircraft, running on hydrogen.

Airbus's scheme, ZEROe, will see them investing in a series of alternative fuel systems and aerodynamic configurations, assembled around three aircraft concepts. The first, a turbo-prop design, would carry up to 100 passengers for 1,000 nautical miles on short-haul trips. The second would carry between 120 and 200 passengers more than 2,000 nautical miles using turbofan engines. Finally, there's a turbofan plane with an exceptionally wide body that blends into the plane's wings, with the same passenger number and range. Airbus is committed to undertaking a first flight by 2025.

The French are not alone in turning the recent retrenchment of the aviation industry into an opportunity for reform. Norway has mandated that 0.5% of aviation fuel in the country must be sustainable this year, growing to 30% by 2030. It wants all short-haul flights to be 100% electric by 2040. Canada implemented a carbon tax on air fuel for domestic travel of 30 Canadian dollars (around US $21) per tonne of CO_2 in most of its regions. Meanwhile, the UK government's Jet Zero Council brings together Rolls-Royce, Airbus, Heathrow, the International Airlines Group and Shell to help the aviation sector make a green recovery. All these developments should be good news for hydrogen in the sky.

How much is this all going to cost? Well, bigger planes are likely to cost more. Cryogenic systems will be an expensive novelty to inspect and maintain, at least for a while. And because of the low density, refuelling a hydrogen aircraft is going to take longer than refuelling a conventional aircraft, even using double the number of fuel hoses. This could mean that each hydrogen aircraft makes between 5 and 10% fewer flights per year than its conventional cousins. The planes, though, are only part of the air travel infrastructure. Hydrogen will transform airports.

Many modes of transport meet at an airport: trains, cars, trucks, buses, aircraft . . . forklifts. Every vehicle in the vicinity may well operate on a fuel cell. This is already happening in Germany, where Munich airport's hydrogen-powered buses have already covered more than 350,000 km. The first public hydrogen filling station was built here.

Safety needs to be prioritised. But airports can become hubs where energy is transformed, stored and distributed.

Regional airports would probably take the lead at first, since they have fewer congestion problems, and finding space for liquefaction plants and liquid storage will prove less difficult than at larger, busier airports. To meet the low level of demand of short-range aircraft, liquid blue hydrogen could simply be transferred by tankers from nearby factories.

As demand rises, it will eventually become economic to pipe gaseous hydrogen directly to airports, which will have their own liquefaction plants. While some new hydrogen pipelines will be needed, many existing natural gas pipelines could be rehabilitated.

Green hydrogen will probably be a cheaper option for airports serving areas rich in renewables, such as those in mountainous areas with plentiful hydropower, those near the North Sea coast with access to wind power, and those in southern Europe, south-western US and Australia with reliable solar. These airports can build their own electrolysis plants.

With a mature supply infrastructure, airports can become energy hubs, lit and heated through hydrogen boilers and fuel cells, and even supplying energy to nearby industries.

None of this need concern you, the future air traveller. Indeed, the rise of hydrogen means that you can be happily unconcerned by other things too: air pollution and the fate of the planet. If your job requires monthly meetings on another continent, no problem. We can all take a guilt-free weekend city break or visit our families even if they live half a world away. We can finally regain the joy of flight.

19

IT IS ROCKET SCIENCE

Hydrogen is a ferociously energetic rocket fuel. It has already enabled us to venture into space. In new engines and new forms, it could soon propel us further and faster than ever before. It could even help us get to Mars.

Powering rockets won't solve climate change. But no discussion of hydrogen's merits as an energy carrier would be complete without the story of how it has helped push the limits of what was possible in space exploration.

This story begins before the Wright Brothers took to the skies, when Konstantin Tsiolkovsky, a Russian mathematics teacher, set about the calculations necessary to get us into space. Tsiolkovsky was born in 1857 in the village of Ijevskoe in western Russia. When he was eight, his mother showed him a nitrocellulose balloon filled with hydrogen. At fourteen, he tried, but failed, to make a paper version. He later developed the idea of a metal dirigible and published papers on the idea. Not bad for a youth whose deafness had forced him out of formal education at the age of ten.

Eventually, Tsiolkovsky settled into a rewarding career, teaching mathematics in the provinces. In his spare time, though, he was designing space rockets and steerable rocket engines, space stations and colonies, airlocks and closed-loop life-support systems. He also (perhaps not surprisingly) wrote a few science-fiction novels.

Tsiolkovsky knew that to remain in orbit around Earth, a vehicle would have to reach a speed of at least 8 km a second. This was easy to calculate, based on Newton's laws of motion. The impressive thing is that Tsiolkovsky worked out the theory of how to reach such speeds using a rocket – a vehicle propelled by ejecting mass out of its rear.

Rockets have to carry all that propellant, which makes them sluggish to start with, but then able to accelerate faster and faster as the burden of propellant decreases. Tsiolkovsky developed an equation that tells you the speed achieved by a rocket, given its starting and ending mass, and the exhaust velocity – how fast the propellant is hurled out behind it. From this equation, he understood as early as 1903 that a single rocket would probably not suffice to reach orbit: better to stack rockets one on top of another. So multistage launchers were another of Tsiolkovsky's inventions.

It makes sense for the propellant to also generate the energy that forces it out of the rocket nozzle – usually by combusting a fuel and oxidiser. Barely five years after the Scottish chemist James Dewar had first liquefied hydrogen, Tsiolkovsky contended that liquid oxygen and liquid hydrogen were the best combination for a space rocket. They would carry the highest energy by mass, producing the highest exhaust velocity.

LH2/LOX propulsion really took off in the 1950s. In 1956, the US Air Force put resources into a project to build a hydrogen-fuelled aeroplane, and though the project itself was eventually abandoned, its managers, technology, liquefi-ers, and other equipment all ended up contributing the upper-stage rocket for the *Atlas-Centaur*, the first hydrogen-fuelled rocket that flew. The Air Force, the army, NASA and the Advanced Research Projects Agency (ARPA), were all working on a large launch vehicle – a scattered effort brought together at last when NASA was given the job of delivering a rocket that could lift a moonshot into space.

The layout of the different stages of the *Saturn 1* neatly demonstrates the trade-offs that designers have to make as they juggle hydrogen's power and its volume. It was clear from early on that LH2/LOX couldn't work to lift the whole rocket from the launchpad. Liquid hydrogen's low density means it needs big fuel tanks, which add weight; and because they make the rocket larger, they also increase atmospheric drag. Consequently, a rocket's main stage is usually powered by compact kerosene and liquid oxygen; LH2/LOX is reserved for the upper stages of a rocket, to be used where the atmos-phere is thin or altogether absent.

One other drawback to LH2/LOX is worth mentioning here, since we encounter the same problem in more down-to-earth environments. That's the effort required to keep hydro-gen and oxygen in their liquid states prior to launch. Seconds after a rocket is filled with these fuels, they reach boiling point. The vapour we see venting from the sides of a rocket's main stage is usually liquid oxygen, but hydrogen's rate of boil-off is actually far worse: roughly half of the liquid hydrogen purchased to fuel the Space Shuttle's three main engines was lost in this way.

LH2/LOX is highly explosive and tricky to handle, but after years of frustration and some spectacular explosions, a rocket with a liquid-hydrogen fuel upper stage rose from NASA's Cape Canaveral Launch Complex on 27 November 1963. LH2/LOX-fuelled Centaur stages, mounted on Atlases and Titans, went on to propel most NASA space vehicles into orbit and beyond. The massive *Saturn V* vehicles used liquid hydrogen stages to launch men to the moon in the *Apollo* programme, and were also used for the Skylab missions in the 1970s. The three engines on the Space Shuttle orbiters burned as much as 230,000 kg of hydrogen every time they flew.

No one has yet managed to propel a vehicle into orbit using just hydrogen and oxygen, but a British project promises to bring that achievement closer. UK company Reaction Engines is developing a new type of engine called SABRE (synergetic air-breathing rocket engine). This is neither a conventional rocket engine nor a conventional jet engine, but a hybrid. During the initial part of its ascent to space, SABRE scoops up atmospheric air and burns it with liquid hydrogen. At an altitude of about 25 km and travelling at just over five times the speed of sound, the engine then switches to pure rocket mode, burning liquid oxygen and liquid hydrogen from onboard fuel tanks for its final climb to orbit. Using such an engine, a spaceplane with the same mass as current launchers could carry twice as much payload to orbit, after leaving the atmosphere on a steep climb.

Assuming it survives the development process, this could also transform long-haul air travel. Scimitar, an engine derived from the SABRE concept, is being designed to propel a radically designed aeroplane, called the A2, from Brussels to Sydney in just over four-and-a-half hours, a journey that takes a whole day in a normal aircraft.

A far more radical possibility would be to use metallic hydrogen. Squeezed under a pressure of millions of atmospheres, the

H_2 molecules are expected to break into separate protons and free-swimming electrons. This could be the most powerful chemical rocket fuel in existence, more than three times as powerful as liquid H_2. Even better, metallic hydrogen would be around ten times as dense as liquid hydrogen, making for more streamlined rockets and bigger payloads, and potentially handy for air travel and other applications. But this is for the far future, and it will only be possible if hydrogen stays in its metallic state after you have released the extreme pressure, which is still unknown.

While rocket engines are a dazzling demonstration of hydrogen's explosive power, this versatile gas is also enabling space exploration in a quieter way, through fuel cells.

Starting in the mid-1960s, NASA's space programme used alkaline fuel cells to generate power for satellites and crewed capsules – providing onboard electricity, heat and water for both *Gemini* and *Apollo* astronauts.* Liquid hydrogen and liquid oxygen were stored in cryogenic tanks, providing the fuel for an alkaline fuel cell that supplied electrical power for both the *Apollo* Command and Service Modules. As a by-product it produced half a litre of water an hour, which was used for washing, reconstituting dried food, and drinking. In the early version, quite a lot of hydrogen was dissolved in this water, and it gave the three astronauts of *Apollo 11* bad stomach cramps. Loud complaints were relayed to NASA Mission Control and happily, the

---

* When welcoming their inventor Thomas Bacon to the White House, President Richard Nixon told him: 'Without you Tom, we wouldn't have gotten to the moon'.

problem was solved in time for *Apollo 12*. Astronauts aboard the International Space Station (ISS) get their drinking water the same way today.

You might think outer space would be cold enough to keep liquefied gases in their liquid state. But space is empty, so heat can't be lost through convection or conduction, the way one's body heat is whipped away by a breeze or a cold bench. Meanwhile, heat comes from sunlight and onboard activities. So carrying liquid hydrogen for a long time on a space mission is asking for trouble. *Apollo* missions were rather short, lasting about a week. If we want to rely on fuel-cell technology for longer periods – for, say, flying to Mars – then cryogenic storage isn't a safe option.

Future long-range missions will get their energy from solar panels, but they will still use fuel cells. One reason is backup power. When a spacecraft reaches Mars and goes into orbit, it will periodically pass through the planet's shadow, and you don't want power to keep cutting out. The best solution is a reversible fuel cell, which can take energy from solar panels to turn water back into oxygen and hydrogen through electrolysis. This is going to be a complex job, but the weight savings over batteries make it a clear choice.

Reversible fuel cells also bring life-support benefits. Astronauts need to breathe, and on the ISS they get their oxygen from water, using electrolysis. The hydrogen, at least until recently, was considered a waste gas and was simply vented into space. As astronauts breathe, they exhale carbon dioxide, and this too must be removed from the air and discarded. Large amounts of water have to be transported regularly to the ISS to keep these processes going – a luxury that future long-duration missions will have to do without.

So now the ISS is making water from exhaled carbon dioxide and hydrogen. Half the hydrogen in the process ends up as a waste gas, methane. But it means cargo ships only need to carry half the amount of water to the ISS, plus a little hydrogen. In the future, the station could become more self-sufficient by using methane cracking, establishing a closed cycle between water, oxygen and carbon dioxide, so not a molecule is wasted. This kind of hydrogen technology is bringing that much talked about first crewed mission to Mars closer to reality.

If we're ever to be serious about exploring and even colonising space, then we need to find a way of creating fuel wherever we happen to visit, from local materials. The relatively recent discovery that water is everywhere in the solar system, from the moon's polar caps to Pluto, is extremely encouraging. One fairly conservative NASA estimate suggests that there might be 600 million metric tons of lunar ice for us to harvest. And where there's water, remember, there's rocket fuel. Filling rockets with liquid oxygen and liquid hydrogen, made by splitting lunar ice through electrolysis, sounds a little far-fetched, but don't be so sure: last year NASA's legal team proposed the legal accords necessary for the world to begin mining the moon for ice.

Harnessing the stuff of Mars for fuel is a little more complicated, but the chemistry involved is nothing we haven't already discussed in this book. The atmosphere on Mars, which has around 0.6% the pressure of Earth's, consists almost entirely of carbon dioxide. This is good news for space travellers, because the carbon from carbon dioxide, plus hydrogen, makes methane. Save the oxygen, and use it to burn the methane, and you've got yourself an ideal rocket fuel for the low-gravity, low-pressure environs of the Red Planet.

The missing ingredient there is hydrogen. At least, we thought it was missing. On 31 July 2008, however, NASA's *Phoenix* lander confirmed the presence of water ice on Mars. We're still not entirely sure how much ice we'll have to play with once we get there. One day it may be possible to extract hydrogen from Martian water ice using electrolysis, making rocket fuel entirely from Martian materials, but we shouldn't stake our lives on this idea just yet. That's why SpaceX's plan for a manned mission to Mars later this century involves sending an initial unmanned mission, carrying a payload of hydrogen. The first crew will be able to manufacture fuel for their return using this hydrogen and atmospheric CO_2. While they're on the Martian surface, they'll look for water that subsequent missions can use to make their own hydrogen on site, giving our universal energy interconnector a foothold on another planet.

Hydrogen's potential to enable space exploration is hardly related to the main scope of this book. But it does show how the tools and technologies we are developing to tackle global warming will give rise to fascinating new ideas and opportunities.

20

SAFETY FIRST

Safety has to be our top priority as we find more and more uses for hydrogen. We need to be clear about what the risks are and put in place strict global standards. There is no room for mistakes, because even small accidents will cause long-lasting damage to this nascent industry.

In a questionnaire passed around the citizens of Munich in 1997, people. were asked what they associated with the word 'hydrogen'.[1] One might think the *Hindenburg* disaster would loom large in people's consciousness. It was, after all, a terrible accident of early aviation, involving industrial quantities of hydrogen and vividly captured on film. In fact, the largest risk associated with hydrogen turned out to be the H-bomb. Nearly 13% of people, confronted by the word *hydrogen*, conjured up nuclear Armageddon, which of course has nothing to do with the chemical reactions that occur when using hydrogen as a fuel.* Yet those

---

* The H-bomb is based on nuclear fusion, the process that powers the sun. This generates millions of times the energy of the chemical reactions we are talking about in this book, and it is only achieved under extreme temperature and pressure many thousands of times higher than anything that will ever happen to your car. Incidentally, the H-bomb is a bit of a misnomer: it

same people, returning home, at some point in the evening cooked supper, perhaps on a gas range, in a room containing a fridge. A third of fridges contain isobutane as their refrigerant, a hydrocarbon that is both flammable and explosive. I doubt a single person cared about that. I know I wouldn't have.

We are constantly measuring our environmental risks, but we do so in a fairly ad hoc fashion. We have no choice, given the complexity of the world and the busyness of our lives. We rely on what we happen to hear, and rely more than we perhaps realise, on the wisdom of crowds.

In 1998, the word *hydrogen* would have hardly ever come up in ordinary conversation except in the context of the Cold War. As for the risks inherent in cooking on a naked flame next to a fridge, they are so minuscule, I doubt anyone outside the white-goods industry ever gives them a second thought.

Those of us working towards a hydrogen future have two battles to fight. We need to make hydrogen as safe as possible. And we have to explain hydrogen to people who know very little about it and have no time to spend worrying about its safety, any more than I have time to spend lying awake fretting about my fridge.

In May 2019, a hydrogen storage tank exploded at a government research project in the rural South Korean city of Gangneung. It killed two people and injured six more. A preliminary investigation found that the blast was caused by a spark, after oxygen found its way into the tank.

The very next month, there was an explosion at one of

---

doesn't use hydrogen, but rather deuterium and tritium, hydrogen's heavier cousins.

Norway's three hydrogen fuel stations. Thankfully, nobody was seriously hurt, although two people received minor injuries when the airbags in their cars deployed, presumably triggered by the blast.

At time of writing, these were the worst hydrogen-fuel accidents of the modern era. Like any technology, hydrogen is not and never will be completely safe. It burns, and its flame, which is invisible, propagates very quickly. Like natural gas, it is odourless. Fossil fuels, too, have residual risk. Coal causes over twenty-four deaths per TWh of energy produced, oil over eighteen deaths and natural gas just under three deaths. These figures, from an article in the *Lancet* in 2007, reflect the risks in producing the energy, not in using it.[2] But we have learned a lot from our experience in handling fossil fuels that is directly applicable to hydrogen. The oil and gas industry has had some very serious accidents, and if you look at why they happened there is almost always an element of human error, or a safety procedure that was missing. One way companies try to prevent this from happening is by running safety audits, looking in particular at accidents that almost happened (called near misses) to devise stricter and better safety standards. They also try to instil a safety-first culture – many industrial plants have a giant display monitor counting the days since the last incident, however minor.

Hydrogen accidents may well increase as its use becomes more commonplace, simply because there will be more opportunities for things to go wrong. Even so, there is mounting evidence that hydrogen can be made safe to use, perhaps even safer than the fuels we burn daily. For example, hydrogen is less of a fire risk than most other fuels. Unless there's at least 4% by volume of hydrogen in the air, hydrogen cannot be set alight.

By contrast, gasoline is flammable above just 1.4%.[3] If hydrogen leaks into the open air, it rises (being much lighter than air) and disperses, so its concentration rapidly falls below that explosive level.

The main risk with hydrogen, then, is not pipeline leaks into the open air. Instead, the risk is small quantities seeping out into closed rooms. Hydrogen can ignite anywhere up to 94% concentration, giving it the widest flammability range of any fuel. And it doesn't need much of a spark. At its most flammable concentration, of 28%, just 0.02 millijoules of energy is enough to ignite the stuff. That is only 7% of the energy required to ignite natural gas. The rule will be to ensure that spaces are well-ventilated, for example; or, where ventilation is a problem, to avoid sparks and naked flames. In a house fuelled by 100% hydrogen, it may be worth adding ventilation outlets at the highest points of rooms.

Safely switching domestic supply from natural gas to pure hydrogen is a known and manageable challenge, but we should never be complacent: every building and every heating system is subtly different and unforeseen issues are bound to turn up. There will be errors and there will be accidents. We must do all we can to prevent them.

In transport, the priority will be containment. Back in the days when we were using hydrogen for buoyancy, we held it in the lightest possible vessels we could fashion. So many thinly stretched intestines were used for the *Hindenburg*'s hydrogen bags that building the airship triggered a national sausage shortage. Now that we're treating hydrogen as an energetic fuel, sausage skins won't do.

Hydrogen-fuelled planes are being designed with their tanks above the level of the cabin, so as to make it effectively impossible for the hydrogen to come anywhere close to passengers or crew. On the roads, hydrogen tanks used in modern fuel-cell vehicles are made from multiple layers of resin, carbon fibre and fibreglass. The twin hydrogen tanks of the Honda Clarity are made of aluminium and carbon fibre designed to resist both extreme pressure and extreme heat. The Toyota Mirai boasts triple-layer hydrogen tanks capable, the company says, of absorbing five times as much crash energy as a steel petrol tank. Toyota were so determined to prove their tanks were safe that they took a gun to them. When the first bullets they tried bounced off, they moved to high-calibre armour-piercing rounds, and even then the tank had to be shot in the exact same spot twice before it ruptured.

Say the unthinkable happened, and your hydrogen tank did rupture: what then? Just as with a pipeline, the hydrogen gas would rapidly dissipate into the atmosphere. And the high pressure in the tank, which sounds frightening, is actually useful – it means no oxygen can get inside, so even if the car or the leaking hydrogen catches fire, the tank won't explode. By the time pressures have equalised, there is hardly any hydrogen left in the tank, certainly not enough for a dangerous explosion. In contrast, if a petrol tank is punctured, the situation is more dangerous. Gasoline is a highly flammable fuel, and one that does not escape like hydrogen. Instead, it leaks out and pools beneath the vehicle, creating a ready source of fuel for a prolonged burn. So, while there are inherent dangers with any combustible fuel, hydrogen can be safer than gasoline. Bear in mind that we already know how to handle hydrogen safely enough that hydrogen-powered buses can be found on the busiest streets of our capital cities.

It's a lot to take in. And, frankly, many won't bother. We'll trust instead in the competence of designers and engineers, the law-abidingness of companies, the oversight of regulators. Despite what you might see in the headlines about a loss of faith in public institutions, the public is not grudging with its trust when it comes to technology.

In 2019, an anonymous social media survey asked two related questions regarding people's views of hydrogen as a safe energy source. Only 49.5% of respondents believed that hydrogen is generally safe, yet 73.2% showed a 'willingness to use hydrogen-powered modes of transportation'.[4] That means about half of the people are wary of hydrogen, which is a reasonable point of view: it is a rocket fuel, after all, but most of these people would still be prepared to board a hydrogen-powered bus. There's no contradiction here. People may not trust hydrogen, but they trust buses. They accept that hydrogen, for all its risks, can be handled safely by conscientious engineers. Now that's a sacred trust, if ever there was one, and it obliges those of us in the industry to be prudent and honest, as we spread what's essentially a rocket fuel to every corner of our lives.

Part 4

Ready for Take-Off

21

THE MISSION

To stave off catastrophe, our global vision for hydrogen must become a reality soon. We have to reach a crucial tipping point in the cost of green hydrogen, making it competitive with fossil fuels for some uses, after which it will grow with a snowball effect. If we act fast, we can make this happen within five years.

On good days, I awake with hope – and excitement – from dreams filled with sunlight and wind, guilt-free flights, fuel-cell buses and green hydrogen boilers. I imagine a world free from the looming anxieties occasioned by runaway climate change. Yes, there will be challenges, controversies and tragedies ahead, all caused by a century's burning of fossil fuels. But, as we know from the Covid-19 pandemic, there is a world of difference between facing a global threat unarmed and having the tools to fight it.

Then there are the bad days. Days when the problems we face seem too many, and the inertia too great to overcome. Our dismal lack of progress makes me feel downcast. We're still burning coal. We're still burning wood. We haven't learned. The things we do in full knowledge of their potential to cause disaster are too many to count.

We do them because we've always done them, because there's no workable alternative, because stopping or freezing our lives and our economies is a terrible sacrifice or because we believe that even if we – personally – stop, everyone else will just carry on. And, a lot of the time, we just do what we've always done without really pausing to think. We know we are headed for catastrophe, but we can't quite believe it – not while today looks much the same as yesterday and there are emails to answer, family and friends to hang out with and dinner to look forward to. Life goes on with its mix of drama, complexity and joy. Some people have already given up and have shifted from looking for solutions to thinking about how to adapt.

So the question is: can we turn all this around, and hit net zero within thirty short years? I now believe we can. And the thing that most gives me hope is the falling cost of renewables and hydrogen. It gives us a vision of what we are trying to build. No longer are we talking about 'reducing our badness' or stopping everything. No longer are we having to choose between our jobs and the future of humanity. We now know what a fully decarbonised world looks like. And it's actually pretty sweet.

It is a world of abundance. We've got all the clean energy we need, thanks to an unprecedented scale up in renewable power production and improvements in efficiency. There isn't just plenty of energy, it also gets cheaper and cheaper the more we install. That's because the renewable sector is the exact opposite of traditional resources. With coal, oil and gas, we tend to drill for the easiest and closest reserves first, and then as demand rises or fields are exhausted we gradually explore at greater depths, and for types of oil or gas that are harder to get out of the ground. That means that as we use more and more energy, its production cost goes up.

Sunlight, on the other hand, is always free; the marginal cost is zero. Meanwhile, the more panels we make, the less they cost. And when we have many solar panel factories up and running and the market matures, producers will start competing on costs and technology and prices will continue to fall.*

To get to this point, we will have invested several trillion dollars. Panels and wind turbines, electricity grids and batteries to electrify everything we can, and electrolysers, pipelines and fuel cells for hydrogen for the hard-to-abate sectors. But despite the trillions spent, the power of free sunlight is such that a net-zero energy system is likely to eventually cost less than the one we have now. Moreover, the increased business activity from these investments should have positive effects on growth† and job creation. Of course, if we don't make the investment, we will lose trillions at the very least, with homes, businesses and livelihoods hit by fire and flooding and other consequences of unrestrained climate change. And we could lose everything.

Our net zero future sees us using our cheap renewable electricity directly wherever possible – that's what's going to power our zippy EVs and our newbuild homes. And we will use it

---

* This is what happened in China in 2011, when Europe reduced subsidies. Demand for new panels collapsed and manufacturing overcapacity led to fierce price competition and some manufacturers going out of business.
† According to IRENA, the energy transition stimulates economic activity additional to the growth that could be expected under a 'business as usual' approach. The cumulative gain through increased GDP from 2018 until 2050 would amount to $52 trillion.

indirectly, as hydrogen, to get into the corners of the energy system that electricity cannot reach. To get all the renewable energy we need, we're going to be putting panels and turbines down increasingly far from our homes. Branching out – which for Europe will mean the North African or Middle Eastern desert or the North Sea – is essential if we are to reach full decarbonisation.

Sunlight is best transported as hydrogen in existing infrastructure, which costs just one eighth of the cost of building a power line.[1] And given that we will need a lot of hydrogen to decarbonise our hard to abate sectors, this is a convenient solution.

The switch to renewable energy is already underway. We already have 2,800 GW of renewable capacity, of which 1,500 GW is solar and wind (including 240 GW of new capacity added in 2020).[2] Costs are already so low that renewable power looks competitive with oil and gas (especially if we don't need to make too many big investments in grids and batteries to put it to good use). But what's still missing from this energy revolution is the part that will deliver us the rest of our fully decarbonised world: hydrogen. To reach net zero, we need hydrogen to provide about quarter of the total energy we use.[3]

So far, hydrogen in the energy system is small beer. All the electrolysers that we have installed, cumulatively, amount to a few hundred MW. To put it in context, they provide enough hydrogen to power roughly 7,000 buses. We are looking to raise that amount dramatically, mobilising global investments of $11 trillion in production, storage and infrastructure. And all this from a virtually standing start. How?

The two-dollar tipping point

If hydrogen is to realise its full potential it needs to be ample, cheap, easy to transport, store and distribute – and we also need plenty of hydrogen trains, trucks, steel mills and boilers that can make the most of its precious energy. Unfortunately, we are not there yet. Hydrogen is stuck in the familiar 'chicken and egg' predicament, where supply waits for demand, and demand waits for supply. There's not much hydrogen being produced, so costs are high, so there is little demand, and no infrastructure to connect the two. To break out of this, hydrogen must become cheap enough to compete with fossil fuels, at least in some applications. What would that take?

Well, let's take a look at where we are today. Green hydrogen from electrolysis costs about $5 per kilogram ($125/MWh) in areas of the world where renewables are abundant. Blue hydrogen, from fossil fuels and CCS, is much cheaper at about $2.5/kg ($60/MWh) but there is barely any around because carbon capture and storage projects are few and far between. By far the cheapest option is, of course, highly polluting grey hydrogen, produced by coal or steam reformation at around $2/kg ($50/MWh).

These are just the production costs. The next step is getting hydrogen from where it is produced to where you want to use it. Pipelines that now carry methane are by and large technically ready to switch to hydrogen, but there needs to be a reasonable amount of hydrogen to carry before it makes sense to do so.* Other infrastructure, including storage facilities and

* That's partly because the hydrogen needs to pay for the cost of converting and maintaining the pipeline, but also because hydrogen consumption needs to start replacing methane to free up capacity in the pipelines.

filling stations, needs to be built. And steel mills, freight companies and other potential customers need to be convinced that H_2 can be delivered at low costs and at scale before they make the investments required to switch from coal or oil products.

How much that adds to your bill (the overall premium that you'd need to pay to switch, compared to your current fuel) depends on what you need to invest to transport and store hydrogen, and how much you are planning to use. To start with, when a small amount of hydrogen has to pay for a lot of infrastructure, that can push hydrogen from merely expensive to eye-watering. At scale, pipeline transport should only add between €0.10 and €0.20 to the cost of a kilogram of hydrogen if it is transported for 1,000 km.

The combination of high production costs and low volumes means that at the pump, today, the price of hydrogen can be as high as \$12/kg, or \$300/MWh (for comparison, diesel at US pump prices costs \$70/MWh). No wonder, then, that people are sceptical about switching their energy supply to hydrogen (I have this conversation a lot). They need to see the cost fall substantially.

How substantially? That depends what fuel it would substitute and what we think will happen to the cost of that fuel over time, including the price we put on the CO_2 that it would emit.* The rough calculation on the next page assumes moderately rising fossil fuel costs and CO_2 prices in the coming years. Of course, this is just one possible outlook. If CO_2 costs were

---

* And also whether you need to pay for delivery and storage for your own use, and whether a centralised pipeline and storage system has already been developed.

to rise faster than expected, that would give hydrogen a leg up. Conversely, should a lack of demand for fossil fuels drive their price lower, hydrogen would have to race even further down the cost curve to catch up. Nonetheless, this sliding scale of switching costs is a useful guide.

If you run a railway company, hydrogen is not far off the mark, even at **$5/kg** production cost. Trains are among the first sectors in which hydrogen becomes competitive, because they take fixed journeys, so you only need one refuelling station for a lot of trains.

If you run a trucking company, you will probably need hydrogen production costs to fall further. That's because trucks need to have more filling stations than trains. To be cost-competitive with diesel, hydrogen would need to hit around **$3/kg** ($75/MWh). This will open a big market. Just in Europe, the US and China it would total 4,000 TWh, or 100 million tonnes of hydrogen. But reaching this market will be a relatively slow process, requiring a lot of additional infrastructure.

For green hydrogen to make sense for many big industrial uses, it needs to more than halve to **$2/kg** ($50/MWh). That's when it becomes competitive with grey hydrogen and can be used as feedstock for ammonia production and in refineries. This opens a 70-million-tonne market worth more than $130 billion a year, which can be reached quickly as the market already exists.

To compete with natural gas in heating and coal for industry, in most countries we need to get somewhere below **$1/kg** ($25/MWh), at which level hydrogen begins to displace fossil fuels in many sectors around the world.

For some of the hardest-to-abate sectors, like shipping and

aviation, it will be possible for hydrogen to compete with fossil fuels only if society imposes a significant additional cost on CO_2.[4]

This price scale means that clean hydrogen's growth should start to speed up when it gets to \$3/kg, and it will reach a tipping point around \$2/kg where it becomes cost-competitive in existing markets.

Getting hydrogen to \$2/kg is our mission.

The path to the tipping point

When someone is trying to sell me something and tells me that its costs will need to fall by more than 50% before any big consumers will want it, I tend to stop listening. I assume the process will take forever, and that whatever I am being sold will be overtaken by events. That's what most of my potential clients think, too, when they work out the cost gap between green hydrogen today and the fuel they are using. But the next bit of the conversation is where I see their eyes light up, what we call the 'holy shit' moment. And that's when I show them just how easy it will actually be to halve the cost of hydrogen.

Today, when we spend \$5 on a kilogram of hydrogen, about \$3 of that cost pays for the renewable electricity to make it, while \$2 pays for the electrolyser that we had to buy to turn water into hydrogen. There is also an inbuilt assumption about the return that whoever has built the renewables and electrolyser needs to make on their money, every year, to want to do it. All three of these costs are on their way down, fast.

The cost of the renewable power we need to produce

hydrogen is plummeting, from \$1,000/MWh twenty years ago to as low as \$10 today. And every forecast we make about where it will get to next year is always too conservative.

The falling cost of electrolysers is why this book exists, and we will get to it in a moment. But before we do, consider the impact of the cost of capital; the return that people require in order to risk their hard-earned cash on hydrogen projects. Because such projects are still largely experimental, today the expected return is something like 8–10%. But these risk calculations are becoming more favourable as more investors have realised that getting involved in the energy transition now, rather than later, is the better strategy. Renewable infrastructure has an essential role to play in humanity's long-term future, and there is a mountain of green capital chasing opportunities.

The risk involved is also coming down. Some of these new hydrogen investments will be backed by government policies that help to guarantee their returns, and this is likely to happen more often now that governments are aiming to 'build back better' after the Covid-19 pandemic. This will mean hydrogen incentives – like those in renewable power before them – will not be subject to the kinds of wild price fluctuations that dog fossil fuels, and this reduction in risk serves to lower the cost of capital.

Hydrogen investments are much easier to develop and run, too. We don't run the risk of spending millions of dollars to drill wells hundreds of metres deep, and then find out our modelling is wrong and there's nothing there after all. That happened to me a few times and isn't pleasant. Every oil and gas project is an engineering feat because each field is different, invisible to the naked eye and unpredictable. By contrast, hydrogen and renewables are above ground, and they are about manufacturing, which is standardised and predictable. Finally, geopolitical risk can be much lower because hydrogen production is not nearly so restricted, geographically.

These three factors – stability, standardisation and ubiquity – lower the risk for investors to the point where they don't need to demand the kinds of high returns we're used to seeing in the fossil fuel industry.

Traditionally, projects to explore for and produce oil and gas have looked for returns well above 10%; indeed, many argue that today, investors should be demanding even higher returns, when so many of their peers seem inclined to divest from fossil fuels, and when risks associated with stranded assets and CO_2 costs can only rise. Auctions of renewables are quite different; they're won at a 5% return on capital. We expect hydrogen to be somewhere in that ballpark.

Lots of people can see what's happening in the financial markets, and to the cost of renewables. But it is the other ingredient – the cost of electrolysers – where the prospect for massive cost reductions is not widely known. Electrolysers are cheaper than they were, but much, much dearer than they should be. And that's because we make so few of them. All the electrolyser capacity ever installed in the world is something like a few hundred MW, peanuts compared with the 130 GW of solar capacity we installed in 2020. The small scale of the business means that they are still essentially handmade. I've spoken to manufacturers – and invested in two companies[5] – and it turns out some factories are only making a few big electrolysers per month, with some of the work being done by hand.

If Poul la Cour, the Danish wind power pioneer we met in

Chapter 6, could see us today, I think he would be dismayed at the lack of progress over the last 200 years. Especially since there's nothing intrinsically very expensive in electrolysers. Their cost should come down rapidly once economies of scale get to work. Automated production of the electrolyser components will bring down the cost for the electrolyser stacks, and scale will also reduce the costs of things like compressors, gas cleaning, demineralised water production, transformers and installation. As production scales up, so will the volume for suppliers, which will further drive down the cost of the finished product. We are looking for a trajectory similar to that of the large-screen TV. The first 42-inch plasma TVs sold in the US for $15,000. You can buy the LED equivalent today for $300.

There's a way to model this process called the learning rate. This is the percentage fall in cost that happens when you double installed capacity. The learning rate shows us which technologies are likely to become significantly more accessible in the future, and which are mature, with all their big surprises behind them. Onshore wind turbines for example, have been improving at a 12% learning rate, while photovoltaic technology has achieved a stunning 24%.

What about electrolysers? An analysis by *Bloomberg* concluded that the learning rate for alkaline electrolysers is about 18%, and for less mature PEM electrolysers it's even better, at 20%.[6] Thus, if we double the capacity of all the electrolysers in the world (not a hard ask at the moment), their cost should fall by nearly 20%.

Eventually, technologies become mature and we get to a cost level beyond which further reductions are difficult to achieve. Where might that be for electrolysers? We interviewed manufacturers during a hydrogen event that we held in Rome, the Hy-Challenge (the world of hydrogen is full of terrible puns). And it turns out that, from today's reference cost of almost $1,000 per kW in Europe, some firms reckon they could build electrolysers for $150 per kW if demand were high enough, while some credible long-term estimates are as low as $130 per kW.[7]

The chart on the next page shows how the different cost elements interact. If you take mainstream projections for the declining cost of renewables, determine how many electrolysers you think will be coming onstream in coming years and apply a plausible learning rate to them, and then apply a cost of capital, you get a forecast of the cost of green hydrogen over time. Using the same model, you can back-calculate the capacity needed to get to the $2/kg tipping point.

This table was originally presented in a study we published in 2019, and every year we check it against market trends and find that things are moving faster than expected. It's proved so useful that it has, on occasion, been presented back to me, around the world and in different languages.

For the cost of renewables, we used publicly available cost curves, such as those provided by Bloomberg. We put the electrolyser learning rate at a conservative 15% instead of the 18–20% predicted by Bloomberg. Plugging that into the model, we worked out that we would need only 25 GW of capacity in the next five years for the cost of hydrogen to fall to between $2 and $3 per kilo in many areas of the world. In the sunniest and windiest regions, it would be below $2 per kilo.

This may sound like a stretch but if you think that the world's lowest-cost solar auction was won at $10.4/MWh, and that the cheapest electrolysers in the world may already be available for

$200/kW in China,[8] green hydrogen could already theoretically cost as little as $1.5/kg.

Table 6: Hydrogen cost decline by component

Year	$/MWh renewable cost	Electrolyser capacity implied GW	Electrolyser Capital expenditure $/kW	Cost of hydrogen $/MWh	Cost of hydrogen $/kg
2010	360	n/a	1500	$600	24
Today	30–45	0.3	950	$100–140	4–5.5
+ 5 years	20–35	25	330	$45–70	2–3
+ 10 years	15–27	50	270	$35–55	1.5–2
Large-scale adoption	10–13	>50	170	$22–28	<1

Forecasting is an art, rather than a science, so getting hung up on the precise number isn't terribly useful. Let's say something in the region of 25 GW is how much electrolyser capacity we need to build, across the world, to get hydrogen costs down to $2 per kilogram in some parts of the world, the level at which it becomes an economically appealing fuel to pump into your truck, train or fertiliser plant.*

And if you think getting green hydrogen from $5 to $2 per kilogram is ambitious, just think that Nel reckons hydrogen will cost $1.5 per kilogram by 2025, and the Department of Energy in the United States launched an 'Energy Earthshot' initiative in June 2021 aiming to get clean hydrogen to $1 per kilogram by 2030.

We need policy support and incentives to get us to this point, but only for a short time. Then free-market economics

---

* 25 GW of electrolyser capacity may seem like a lot given hydrogen's starting point today. But to put it in context, the IEA's special report 'Net Zero by 2050' envisages 850 GW of electrolyser capacity by 2030.

will take over. Demand will rise and costs will fall still further, opening up new multi-billion dollar markets: prairies in which hydrogen can run free – without taxes, subsidies or other assistance. In the longer term, hydrogen costs can fall below $1/kg, at which level it will become competitive with coal in some applications.

Of course, these are production costs, and hydrogen's capacity to compete with fossil fuels will also depend on scaling up infrastructure so that it can be transported cheaply and at scale. So it makes sense to focus policy support on that, too.

The three Cs – making it all go faster

Hydrogen makes such good sense that it is almost certain to reach its $2 tipping point eventually. But if we were to rely on market forces alone to push the ball up the hill, that could take a long time – time we don't have. We would have to rely on the niche uses for green hydrogen that already make economic sense, such as forklifts, and on very favourable locations for renewable production. As solar and wind costs fall further, a few more niches will gradually open up, like for instance our old friend the non-electrified train, and the ball would speed up a little, but with such a piecemeal adoption of green hydrogen, we face a long wait before hydrogen reaches its $2 tipping point and comes into mass-market use, likely after 2040.

That's not nearly good enough. While green hydrogen slowly grows, all those hard-to-abate sectors will keep pumping greenhouse gases into the atmosphere, increasing the risk that we hit the wrong kind of tipping point, with catastrophic consequences for the climate.

By moving too slowly, we also run the risk of missing useful windows for hydrogen adoption. Between 2020 and 2030, around 50% of European steel and chemicals production capacity will require reinvestment[9] – meaning that machinery in a lot of aged plants will need to be replaced. If hydrogen isn't available at scale in time, these investments will probably be made in existing technologies, which will then need to either lock in fossil fuel usage for another 15–20 years or need to be thrown out before the end of their useful life, wasting money.

Moving quickly to create demand lowers costs more quickly, and then the whole world can benefit from cost-competitive hydrogen. That's what happened to renewable electricity: everybody benefits from cheap solar and wind power today because a few European countries bankrolled the learning curve, with spectacular results. This front-loading means fewer overall subsidies, lower overall emissions, less risk of stranded assets – in short, more bang for our buck.

So, in sum, we need to hit the accelerator, and boost demand now. What we need is a goal to guide all our efforts. I vote for '$2/kg hydrogen within five years'.

As we now know, that involves building roughly 25 GW of electrolyser capacity. That's 10% of the solar and wind capacity we build every year. And we're talking about doing it over five years, so it really should be manageable! There are three key steps to doing this, involving businesses, governments and the general public, which I am calling the three Cs: Companies, COP and Consumers.

22

CATAPULT COMPANIES

The first movers in hydrogen will be companies. Some are working hard to boost their green credentials. A new global coalition, the Green Hydrogen Catapult, is already acting as matchmaker between demand and supply, but that needs to accelerate. Then we need electrolyser gigafactories to help us on our way.

When I started working, in most parts of the world it was a given that the main purpose of companies was to make money for their shareholders, following US economist Milton Friedman's doctrine. The idea is that by pursuing their own interest, they would also create goods and services that people wanted, employ people, pay taxes, and in general benefit civil society. That's not always the way things pan out. The bankruptcy of Lehman Brothers and the ensuing recession highlighted how the short-term interests of companies – or more often highly-paid company managers – can clash with what is good for society as a whole.

Partly as a result of the soul searching that ensued, there's been a shift in what people think the role of companies should be. No longer are they expected to pursue profit and do good as a result. Now they are expected to pursue a societal purpose first, and make more and 'better' money as a result. The two theories sort of converge – they both hold there is a long-term link between the value a company brings to society and the value that it will create for its owners. But there has been a change of emphasis.

By and large, the reputation of companies has survived the 2020 pandemic. Essential services have been maintained. Working from home has been made possible at a faster rate and more smoothly than anticipated. The vilified pharma sector has delivered vaccines at breakneck speed. Indeed, the 2021 edition of the Edelman barometer of trust, an online survey taken by 33,000 people, reveals that businesses are trusted ahead of governments, the media and even NGOs. Thus, it is fitting that companies are often to be found at the forefront of efforts to solve climate change. And that can be a very good thing, because they bring to the table heft, speed and the capacity to innovate.

Microsoft, Amazon, Apple, Google, Unilever, Ikea and many others lead the world in their respective fields. They want to be the first, the fastest and the best. They have financial fire-power comparable to some national economies. Many of these corporate behemoths are leading with their climate change commitments, too. Consider for example Amazon's 'Climate Pledge' – a commitment to meet the Paris objectives by 2040. As the first signatory of this, Amazon promised to buy 100% renewable energy by 2030, order 100,000 fully electric delivery vehicles and invest $100 million in reforestation projects around the world. Microsoft has also made a very interesting promise. It wants to be carbon negative by 2030 – and by 2050 it aims to have absorbed all the CO_2 it has ever emitted since it was founded in 1975. As they compete to offer their customers products and services that are increasingly green and to attract sustainable investment funds, these companies could become the first bulk-buyers of clean hydrogen, which is the only way to provide 100% green fuel for green trucks, energy for data centres and heat for offices.

As the first movers, these companies would have to buy hydrogen at today's relatively high prices. We're talking the current $5/kg production price, plus maybe the same again in delivery and storage costs. How much of this cost would be passed on to consumers? My guess is very little. Compared with the scale of these companies and the cost of their products and services, it would barely be perceptible.

In the US, Anheuser-Busch is already delivering beer using a fleet of hydrogen-powered trucks.[1] For a company like this, hydrogen is a relatively small part of the business and delivery by hydrogen trucks adds only a tiny cost to a bottle of Budweiser – around half of one cent.* For all the positive PR that buys them, it is a bargain. Thus, if more industry leaders can lead the world on hydrogen, they will pave the way for supply to scale up.

The Catapult

In 2019, at a dinner in Davos with the UK COP26 President,† I pulled out an earlier version of this book, pointing to an earlier version of Table 6. As usual, I was trying to show people

---

* My back-of-envelope calculation, with a truck moving 100,000 cans of beer, and the extra cost of hydrogen being $200–400 per round trip, depending on distance.
† At that time the role was filled by Claire Perry O'Neill.

just how tantalisingly close we were to the hydrogen tipping point, so close it was practically within reach. And I saw that 'holy shit' look on their faces. We agreed that to set the transformation in motion, a moonshot was needed. Cue lots more conversations.

What came out of them was a global coalition, called the Green Hydrogen Catapult. Snam and six other companies* announced the Catapult in London, on 8 December 2020. The vision is to bring together hydrogen consumers, producers and infrastructure companies, to have 25 GW of electrolyser capacity planned by 2026, with a view to reaching $2/kg hydrogen in some areas of the world before the end of the decade. This is an ambitious target, but we've hit targets like this before, and recently too. One of our Catapult partners, ACWA Power, led the race to deliver photovoltaic energy at well below $0.02/kWh in some places.

Part of the vision behind the Catapult is that backing a target makes it that much more likely to happen. Producers gain confidence that their projects will deliver hydrogen at $2/kg, and consumers that this fuel will be available at the required price. Banks become more willing to fund projects because they have confidence that they will become profitable, and governments more likely to support early-stage efforts.

For the Catapult to deliver its target, it will need to engage other companies, with a focus on potential hydrogen buyers like the industry leaders mentioned. Many have already signalled

---

* These are ACWA, CWP Renewables, Envision, Iberdrola, Ørsted, and Yara.

their interest in joining. That's when the matchmaking starts, to forge links between early suppliers and early potential buyers.

Matchmaker

Acting as a matchmaker to create a new market is nothing new for Snam. That's how the Italian natural gas market was created in the 1950s. Before then, lots of big pockets of natural gas had been discovered in Italy, but no one knew what to do with them. Indeed, oil company lore tells of grizzled explorers who, upon hitting gas, would simply bung the hole up in disgust and hope for better luck next time. What was missing was a market, and a way to get the gas to it.

Snam provided both. It created demand, by going to energy-intensive industries and pitching the cheaper, cleaner natural gas alternative. Once it had pulled together enough demand to make transport worthwhile, it built out a pipe to connect them to supply. By 1948, it had the first 257 km of pipeline done, connecting gas fields near Parma to industry in Milan and Lodi. The network reached more than 4,500 km by 1960. In the 1970s and 1980s Snam connected the now well-developed Italian network to gas supplies from Russia and North Africa, which are still being used today. The spread of natural gas helped contribute to Italy's economic miracle, in which per-capita income trebled in real terms between 1950 and 1970, and the country became the world's fifth largest economy by 1990.

With hydrogen, we and other gas companies can do the same thing. As $2/kg comes within sight, we can use the visibility of ample, cheap supply to aggregate demand. We can then switch existing infrastructure to hydrogen. Networks will evolve to carry hydrogen, biomethane and also potentially CO_2. As infrastructure extends across countries and around the world, it should lead to liquid markets, lowering cost and improving supply security for consumers.

Companies can also accelerate green hydrogen by clubbing together on supply. Partnerships between producers can speed up the development of electrolysers. We can learn from the experience of Airbus, born when British, French and German aircraft manufacturers decided to put together projects and resources to compete with Boeing, Lockheed Martin and McDonnell Douglas.

The proposal Snam launched in the *Financial Times* in November 2019, and reiterated at the World Economic Forum in Davos, is to set up a gigafactory of electrolysers in Europe, to scale up production and reduce costs, while creating new jobs. Gigafactories such as these would reduce production costs and reassure consumers that their demand can be met.

The Catapult is already collaborating to accelerate the necessary technology, component manufacturing and construction, so companies can get the ball rolling, creating the first few projects. But companies can't get us where we need to be without a helping hand. There will also be a need for policy support, as well as a lot of money from governments (far more than they are currently contributing). So it is great that the Catapult's announcement has already sparked interest from policymakers.

23

CALLING THE COPs

Governments can support hydrogen with policies to boost demand and supply. Blending some hydrogen into the grid can provide an immediate and flexible market. Some sectors or locations are almost ripe for hydrogen, and could be given a little nudge. An international outlook and a focus on R&D will do the rest.

For some years, global government climate meetings have been pretty disappointing things. After the remarkable success of the Paris Agreement, where nearly every country came to agreement, international consensus frayed at the edges. But that is all about to change. At the time of writing, the next such COP (Conference of the Parties) is due to be held in Glasgow in late 2021. The global context has improved, with America back at the helm of climate diplomacy, and a key target for this and future COPs will be to get a lot more countries, companies and perhaps individuals to sign up to net zero.

As we've seen, net zero offers a positive climate goal. Because hydrogen is instrumental to it, I expect quite a lot of nations to advance strategies to develop hydrogen. And here's the important part: it doesn't have to be everybody. The demand that you need to create to make hydrogen competitive is pretty small in the context of things. A coalition of the willing, made up of nations which have already signalled their commitment to hydrogen, would have enough firepower to make the difference.

What kind of policy support would make the most sense for hydrogen though? Well, as we saw briefly in Chapter 4, many an economist would tell you that a global carbon tax or carbon price would be the best policy instrument – for everything. And it's easy to see why. Different actions to eliminate carbon emissions have different costs per tonne of CO_2 abated. Using solar and wind power for power generation won't be more expensive than using fossils, so the cost of getting rid of that CO_2 is zero or even negative. Other options will cost you more than the current alternative, so there is an implied extra cost that needs to be paid for that CO_2 to be abated. The chart on the following page shows CO_2 abatement cost for various things you might do.

Most of the time, economists think that you should get rid of the cheapest CO_2 first. And that's what would happen if you imposed a global carbon tax. If everywhere in the world had to pay €50 for every tonne of CO_2 emitted,[*] all the things that remove a tonne of CO_2 at less than that price would be worth doing immediately (for example switching from coal to gas for power stations, or from coal, oil and often gas to renewables). In time, the tax could be raised to encourage the more expensive abatement options.

---

[*] That's not a fanciful level. At the time of writing, the EU ETS price was above €50/tonne.

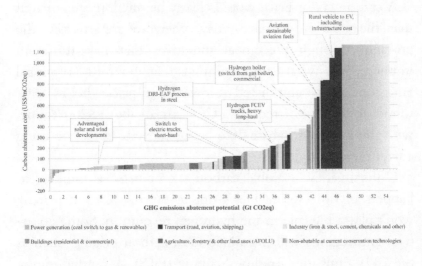

GHG emissions abatement costs
© Goldman Sachs Global Investment Research[1]

The same goes for a carbon price, where companies would be allocated 'permits to pollute' and mandatory reduction trajectories. This also goes by the name 'cap and trade' and allows companies to either overperform on their own targets and sell permits, or underperform on their targets and buy the permits. So every year, they would be faced with the decision of whether and how much to cut CO_2 in their own industrial processes versus buying or selling permits, and they would pick the cheaper option. Every year, they would have to cut a little more, and the price of the permits would rise. This strategy, applied to sulphur dioxide and nitrogen oxide emissions, was how the US solved acid rain in the 1990s – a campaign pursued at the start of his career by Senator Kerry, at the time Lieutenant Governor of Massachusetts.

A carbon tax or price would clearly be much more efficient than the situation we have today, where we are attacking the problem through piecemeal initiatives that have different implied costs. What I find striking is that about 40% of global emissions can be eliminated at low carbon prices, below $50/tonne. Another 40% require a CO_2 cost of above $200/tonne, with the hardest-to-abate 20% being above $700/tonne using current technologies.

Unfortunately, a global carbon tax or price is unlikely to happen anytime soon. That's not to say that carbon pricing won't play a big role in the hydrogen revolution. Some countries and regions will introduce local carbon pricing, such as the EU's Emissions Trading Scheme (ETS). A border adjustment – where importers have to pay a tax equivalent to the carbon content of their goods – may well spur countries that want to export to Europe to introduce their own, parallel, ETS systems. So, over time, we are likely to get a loose sort of carbon market going. By making fossil fuels more expensive, this will gradually help to make hydrogen competitive.

But gradually isn't good enough. We have to speed this process up and reach that tipping point soon, otherwise we won't have the quantities of green hydrogen that we need to switch a quarter of our energy system over by 2050. The least-cost pathway is efficient, but we also need to kick-start the learning process for new technologies that will be essential to reach net zero.

We should also get used to the idea of starting to act before we're completely sure about what the lowest-cost pathway will be. We need to make every effort to work it out, of course, and we should start from the investments which are the most likely to be needed. But if we need to make a judgement call, I would err on the side of action. Worst case scenario, we write off some investments, which, though significant, pales in comparison to the alternative. If we err on the side of inaction, the worst case is runaway climate change.

Countries that want to help push hydrogen costs down – which, as you will remember, requires some 25 GW of electrolyser capacity – should consider six main policies.

Blend

The first is to mandate the blending of green hydrogen in the gas network. This is a way to increase demand immediately without changing infrastructure or waiting for specific consumption sectors to develop. You can develop electrolysers immediately and install hydrogen boilers gradually. Our studies show that 5–10% hydrogen blends can work in existing gas networks.*

This opens up exciting opportunities. For instance, if Europe was to require just 5% green hydrogen in its gas mix, that would by itself increase demand to 35 gigawatts of electrolyser capacity, double what we need to reach our $2 tipping point.

Some reckon that blending is a poor use of hydrogen, because this expensive energy source would go to substitute natural gas, the cheapest and least CO_2-intensive fossil fuel. Yet

---

* In fact, blending doesn't have to be physical. A country can simply impose a mandate on gas suppliers, telling them that a share – call it 10% – of the overall product they sell needs to be low-carbon or renewable gas. It is then up to the gas supplier to find the sector that will pay the most for this decarbonised gas, and that leaves them with the highest profit after all the costs have been taken into account. In general, any scheme that relies on a liquid hydrogen market will require Guarantees of Origin, which certify that the hydrogen really is green (or blue) and act like certificates which can be traded and used to offset carbon prices.

overall costs would be limited by the fact that we don't need to change any infrastructure, so it would just be the additional fuel cost. Being able to blend hydrogen into the network would also mean we could build supply without having to worry about where the first consumers are sited. That would encourage hydrogen production to develop in the most efficient locations, creating a solid foundation for the hydrogen sector.

Blending may only be temporary. You can dial it up or down (as long as it stays within its physical boundaries), which makes the value chain much more flexible. And once we reach a reasonable scale in hydrogen production, we could switch pipelines to carry pure hydrogen and gradually develop those end uses that need it. Or if membranes to separate methane from hydrogen turn out to be affordable, we could continue to blend the gases in the pipeline network and then use membranes to supply each customer with the gas of their choice.

The extra cost involved in our Europe/Japan example would be relatively small – around €8 per person per year. That is similar to the amount drivers in Europe are already paying to have a mandatory share of biofuel added to their petrol. No driver I know is even vaguely aware of this scheme: the cost is so tiny, it vanishes in the fluctuations of the fuel price. Similarly, no one need ever care (though they should certainly be told) that every time they boil a pot of pasta, they are paying an eighth of a cent towards the coming hydrogen revolution.

A blending mandate like this would have dramatic implications, creating enough demand at the stroke of a pen to make hydrogen competitive with fossil fuels in other markets.

Sector support

Another strategy is to go for the low-hanging fruit: those sectors that only need a little financial encouragement to become early hydrogen adopters. That may be because the price differential between their current fuel and hydrogen isn't very high, or because they don't need a lot of infrastructure, or because they need to make lots of investments anyway.

Into this camp falls heavy road transport. For trucks and buses, the cost difference between hydrogen and diesel isn't huge – offset in part by the greater efficiency of fuel cells versus traditional combustion engines. So, a mandate that fuel providers must provide some of their energy from green hydrogen would not be too costly.

What does add to the cost of the intervention is building all the filling stations we'd need to make people feel comfortable with owning a hydrogen truck – and also delivering the first hydrogen to all of these filling stations (remember, we're doing this nudging before a pure hydrogen pipeline network becomes available). It would need to be done smartly, focusing first on buses or truck fleets that either have set routes or spend the night in a depot, and picking areas close to where green hydrogen production can be scaled up, with access to strong renewable potential. But if 10% of European trucking were converted, we would be talking about 25 GW of capacity.

Mandates don't need to be government policy, either – not if the whole sector is governed by the same organisation. That's how global shipping works. And the International Maritime Organization, which runs the show, is keen to burnish its environmental credentials. It has already done a sterling job forcing sulphur out of ship fuels. Now, it could contribute to getting the global hydrogen market off the ground. True, replacing fossil fuels with hydrogen in shipping would be expensive, but we wouldn't need to go for high shares given how big the sector is. If just 1% of the world's ships were powered by hydrogen, it would take 20 gigawatts of capacity to satisfy the market that would be created.

Industry is also angling for either a mandate or an incentive to switch. And here, too, it would make sense to push things along a bit. Replacing some of the world's massive grey hydrogen market with green would be a relatively inexpensive nudge, as all the infrastructure you need is already in place.

If Europe decided to mandate a 10% share of green hydrogen in current grey hydrogen usage, this alone would require an electrolysis capacity of 15 gigawatts. We also wouldn't need every company to have 10% green: one company could go all green and sell its outperformance to everyone else.

A similar logic applies to industries that will need hydrogen to decarbonise, such as steel. Green hydrogen looks very expensive right now compared with the coke traditionally used in in steel production, but it might make sense for policy to bridge this gap.[2] We really don't want anyone to build new blast furnaces, which will either keep polluting or have to be abandoned and written off.

Table 7: Electrolyser capacity needed to supply different use cases

Policy	Hydrogen capacity
5% blend in the European gas grids	35 GW
10% of the current European grey hydrogen market	15 GW
10% of European trucking	25 GW
1% of global shipping	20 GW
European hydrogen strategy	40 GW

Hydrogen valleys

A third policy approach is to create hydrogen valleys, matching up sources of demand and supply physically close to one another.

Choose a large potential consumer such as a refinery or a plant that produces fertilisers and put hydrogen fuel stations nearby. Industrial districts are often already close to ports, so hydrogen could also fuel ships in the same area. By pooling demand in one place, the necessary transport, distribution and storage infrastructure can be optimised. Ports are great places to site hydrogen clusters because they will have access to hydrogen imports, and many industries are already nearby. The Energy Transitions Commission has published this map of promising sites for the first hydrogen clusters.

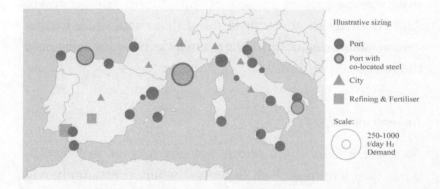

Potential hydrogen clusters in Southern Europe

Urban schemes, such as the one envisaged in the Leeds H21 study, are more complex to implement, but useful for carrying hydrogen close to end consumers. Both require incentives to compensate users for higher fuel costs.

Supply-side support

The renewable revolution was driven by generous subsidies to encourage entrepreneurs to generate renewable power. That gave equipment manufacturers long-term visibility market, so – bit by bit – they built bigger factories and benefited from economies of scale.

Could we have achieved the same result at lower cost if we had just paid manufacturers to scale up equipment production, giving them the money to build huge gigafactories in the first place? After all, the investments in Chinese capacity were in the region of tens of billions. The cost to the European consumer, as we saw, was not far off a trillion. It should be more efficient to scale up supply directly, rather than waiting for the impact of subsidies to trickle down. And countries should bear in mind that establishing hydrogen gigafactories could create a whole new industry, and the jobs to go with it.

Imports and exports

As governments think about policies to give hydrogen a push, they would do well to look at where their supply will come from in the long term. For many it may be convenient or necessary to import hydrogen, while others can export.

Germany's H2 Global policy initiative is a double-sided auction to work this out. Suppliers are asked how much hydrogen they would be able to export to Germany, and its price. Domestic industries are asked how much hydrogen they would need, and the price they would be able to pay. The government matches the lowest-cost supply with the highest-price demand, and bridges the difference.

The European Hydrogen Backbone initiative, formed by twenty-three energy-transmission companies, has looked at potential hydrogen supply and demand areas, and existing gas grids, and figured out what sections could be switched to hydrogen over time. The conclusion is that by 2040 we could have 40,000 km of hydrogen pipelines in Europe, with potential import routes through North Africa, the North Sea, Ukraine and the eastern Mediterranean. EU Vice President Frans Timmermans has talked about the benefits of importing North African hydrogen to Europe. In a speech at the European Parliament in 2019, he said, 'In my dreams, I would create a partnership with Africa, especially in North Africa, where we will build huge capacity to produce solar energy and transform that energy into hydrogen production which can be exported to other parts of the world'.[3]

R&D

The hydrogen revolution isn't banking on any new technologies – just the scale up of those that already exist. However, there is plenty of room for innovation, in production methods, in ways to transport and store hydrogen, in membranes to separate it from methane, and in countless other areas.

Today's electrolysers use rare materials. The sooner we find alternatives, the more sustainable they will be. And because every kilogram of hydrogen requires nine litres of fresh water, the industry will need investment in desalination technologies, which should help to increase the availability and lower the cost of water in many areas of the world.

Rather than letting innovation happen in an incremental fashion, front-loading it through large public-private research programmes would have benefits. R&D expenditure generates immediate jobs and has high payback – if the innovation pays off. It lowers the cost of producing hydrogen more quickly. And if different countries and companies work together on R&D, that could very well give rise to new business partnerships.

International coalitions

I am optimistic that enough countries will focus on the hydrogen opportunity and give us a policy nudge to make hydrogen happen soon. From Europe to China, Saudi to Japan, Chile to Australia, South Korea to Abu Dhabi – many governments are already making commitments through published hydrogen strategies or beginning to take action on hydrogen.

Europe's hydrogen strategy is especially well advanced, part of its roadmap to carbon neutrality in 2050, as laid out in the Green Deal. In January 2021, European Commission president Ursula von der Leyen said in a speech to the Hydrogen Council that 'Clean hydrogen is a perfect means towards our goal of climate neutrality . . . clean hydrogen demonstrates that we can reconcile our economy with the health of our planet.'

The EU Hydrogen Strategy for a Climate-Neutral Europe, published in July 2020, is built around an increase of electrolyser capacity in the EU from today's meagre 60 MW up to 6 GW by 2024, and to 40 GW by 2030 – well above the capacity required.

Europe is doing this by concentrating on applications that can be easily switched to hydrogen. For instance, Germany is focusing on carbon-intensive industries, such as chemicals, steel and cement; Portugal plans to invest €7 billion to turn itself into a hydrogen exporter and plans to blend up to 7% hydrogen into its natural gas supplies.

The total cost of the EU's effort, including new solar and wind generation, transport, distribution, storage and refuelling, is between €320 billion and €460 billion.[4] Governments and the private sector are both expected to contribute, and some funding will come from an investment plan, InvestEU. In addition, the EU's $750 billion Next Generation EU fund will support a green and digital recovery from Covid-19. For anyone with a large-scale hydrogen project in mind, this is a unique opportunity to find funding, and get the ball rolling for the sector as a whole.

South Korea is another keen advocate of hydrogen. Around 6,000 fuel-cell cars were on the roads in 2020, and the country has set a target of 850,000 fuel-cell vehicles by 2030. Calling hydrogen power the 'future bread and butter' of Asia's No. 4 economy, President Moon Jae-in has declared himself an ambassador for the technology, and is overseeing $1.8 billion in central government spending to subsidise car sales and build fuel stations. Car manufacturer Hyundai is doing well out of the arrangement: government subsidies cut the price of its Nexo model by half to about 35 million won ($29,300), and sales of the model surged in response. But Hyundai is pulling its weight, too. Along with its suppliers, it plans to invest $6.5 billion by 2030 on hydrogen R&D and facilities.

Some territories are pursuing specific trading opportunities, and this gives their hydrogen economy a unique character.

The developing relationship between Japan (which has few natural resources of its own, and abandoned nuclear energy following the Fukushima disaster) and Australia (which has more sunshine than it knows what to do with) is, as we've seen, creating a huge industry dedicated to the manufacture of easy-to-transport Australian ammonia for the Japanese market.

Geography and geology have a huge say in how territories develop their hydrogen economies. The most spectacular example I know of is Chile. Already, just under half of Chile's energy is derived from renewables. That figure is expected to rise to 70% by 2030, and to 95% by 2050. Chile's government is confident that green hydrogen will be cost-competitive with grey before 2030. And its confidence is well founded: Chilean green hydrogen potential is huge – about seventy times the current installed capacity. Thanks to its diverse geography, there's capacity for solar in the north, hydrogen in the middle and wind in the south of the country. And thanks to its thin geography, these resources are always close to shipping routes. Never mind looking after its own net zero targets, Chile, a pioneer in bilateral trade agreements on climate, expects to transport hydrogen (as ammonia, most likely) to the US and across the Pacific to Asia.

The United States is not short of natural resources, either, and has the ambition and capacity to become a world leader in green technology. The pressure is on. Early in his administration, Joe Biden received a road map put together by a coalition of major oil and gas, power, automotive, fuel cell and hydrogen companies. Despite years of neglect and four years of institutional climate change denial, the US finds itself already heavily engaged in the hydrogen economy. Solar and wind power are making big strides in Texas and other states. Billions of dollars of public and private investment are pouring into the hydrogen sector every year. Large-scale fuel cells totalling more than 550 MW capacity are installed or planned.[5]

These nations are some of the most promising candidates for an international coalition.

Companies and governments, of course, don't do very well if no one wants to buy their products, invest in them or vote for them. And that's where we come to the third, and most crucial, pillar of our plan. You.

24

THE CONSUMER CAVALRY

There is a lot that we, as individuals, can do to help. For instance, we can choose to buy goods made using green hydrogen, which would add less than 1% to the cost of many products. To choose wisely, however, we will need to know a lot more about the CO_2 content of what we buy and do.

Did you know that for every loaf of bread you eat, half a kilogram of CO_2 goes into the atmosphere?[1] Would you consider spending 1 or 2 % more for your loaf of bread to ensure that it is carbon neutral?

Say there is a ship that carries your next pair of jeans across the ocean. If the ship owners were asked to spend an extra $4 million a year to run the ship on hydrogen, they would probably say no. If the jeans company were asked to pay the extra cost per voyage, they would also say no. If you were asked whether you would pay an extra 30 cents[2] for your next pair of jeans so that shipping could run on hydrogen, you would hopefully say yes. And that would be enough to provide a business case for the jeans company to pay the ship owners to run on hydrogen.

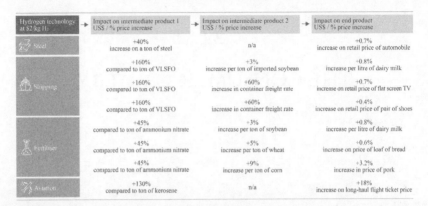

Hydrogen technology at $2/kg H₂	Impact on intermediate product 1 US$ / % price increase	Impact on intermediate product 2 US$ / % price increase	Impact on end product US$ / % price increase
Steel	+40% increase on a ton of steel	n/a	+0.7% increase on retail price of automobile
Shipping	+160% compared to ton of VLSFO	+3% increase per ton of imported soybean	+0.8% increase per litre of dairy milk
	+160% compared to ton of VLSFO	+60% increase in container freight rate	+0.7% increase on retail price of flat screen TV
	+160% compared to ton of VLSFO	+60% increase in container freight rate	+0.4% increase on retail price of pair of shoes
Fertiliser	+45% compared to ton of ammonium nitrate	+3% increase per ton of soybean	+0.8% increase per litre of dairy milk
	+45% compared to ton of ammonium nitrate	+5% increase per ton of wheat	+0.6% increase on price of loaf of bread
	+45% compared to ton of ammonium nitrate	+9% increase per ton of corn	+3.2% increase in price of pork
Aviation	+130% compared to ton of kerosene	n/a	+18% increase on long-haul flight ticket price

NOTE: Calculated for 2 $/kg delivered hydrogen cost.
SOURCE: SYSTEMIQ analysis for the Energy Transitions Commission (2021)

Impact of using clean hydrogen on the price of intermediate and end products

Many make the mistake of thinking that energy policy is only about government and systems and big companies. And, of course, in part it is. It is up to governments to set the rules, and companies to create products. But through the choices you make as a consumer you can steer their work. In 2019, consumer spending was more than $10 trillion in the US – more than the amount spent by the federal government, which was less than $5 trillion over the same period. If we could harness even a small percentage of that to the greater good, it would make a big difference.

Jeans are the least of it. Consumer pressure means car manufacturers want to sell green vehicles, which is already reshaping one of the most difficult-to-change sectors: steel production. And this is just as well, because without customer support, acceptance and eventual demand, it's very hard to see how green hydrogen-based steel will achieve commercial success. By paying a small amount more for decarbonised products, customers will help shift production technologies.

Other manufacturers, too, are increasingly willing to pay a premium for green steel – a movement the European Commission hopes to encourage by a new system of ecolabelling. Labelling would also help appliances: consumers made aware of the emissions from their heating may be keen to buy hydrogen-ready boilers.

The power of the consumer has never been higher than today. To harness it, we need to make CO_2 feel real. People need the information to judge what the best course of action might be. Enough with the gigatonnes of CO_2 emitted and telling the average consumer that they need to get their own annual energy-related emissions down from 4.6 tonnes today to 1.1 tonnes. None of that is terribly helpful, it's too abstract. What *would* help is some sort of a five-a-day labelling system, or something like a calorie counter for CO_2 indicating the percentage of daily allowance that is being emitted by each action or purchase.

Much could be achieved by a handy app on our phones. It could monitor our progress towards net zero by calculating what we emit as we go to work, eat a steak, or buy a loaf of bread, and then offer certified offsetting options, such as planting additional trees or capturing carbon, to buy with a single click. Users might receive badges of honour, which they could share on social media and use to get discounts on products from participating companies, or access to 'net-zero club' restaurants and events.

Then, of course, there is political action. Voting in elections is vital – and the fact that Greta Thunberg and her generation have reached voting age makes me very happy. And regardless of one's wider beliefs and values, so much about politics, ultimately, is local. Does your local bus route operate hydrogen buses? If not, why not? Find out. There's probably a perfectly reasonable answer, which has to do with a lack of infrastructure, funding or knowledge. Write to your local representative, mayor, bus operator, city authority. Ask what can be done? Who's taking an interest? One way to spur political action is to convince people from different parties to adopt a sensible, common course of action. At a very local level, this may not even be that hard, especially if people learn that the cost of cutting out carbon is often a small fraction of a product's price.

25

MAKING HYDROGEN HAPPEN

This book is about how hydrogen puts net zero emissions within our reach – and what we can do to bring it closer.

For most of my twenty years working in the energy industry, I thought we were in a pretty hopeless position. With 70% of people living in developing economies and two billion more set to join us on this planet, I thought that we would need ever-increasing amounts of energy and I couldn't see a way of supplying all that energy without fossil fuels. Our best alternative, renewable power, was still too small and too expensive – not to mention unsuitable to decarbonise a number of sectors.

Since then, renewable power has come of age. A massive scale-up in solar and wind power is already underway, and this will make renewable electricity much cheaper than fossil fuels. This ample and inexpensive renewable power can be used to make hydrogen, as I realised one afternoon in Milan. And hydrogen is a game changer because it gives us a clean molecule to decarbonise those sectors which electricity struggles to reach. As hydrogen becomes more widely used, the cost of the electrolysers used to make it will come down: hydrogen, too, will be cheaper than oil. And that will accelerate demand for hydrogen, which in turn will further accelerate demand for renewables, and so on in a positive feedback loop.

259

As I worked through the implications of this, I could finally see how we could get to net zero without shackling our economies. I understood how renewables and hydrogen could join together to create a seamless energy web in which electrons can be turned into molecules and back again, and how this would enable the power of the deserts and the oceans to be carried over great distances, stored for long periods, used for heat or as a raw material or to produce electricity when and where we need it. In short, I had a tantalising glimpse of a net-zero system that could fuel economic development, create jobs, spark innovation and foster international cooperation and trade.

I am not the first to have a vision of this sort. Scientists – and novelists – have long believed hydrogen to be the fuel of the future. Why should this time be any different?

The answer is that hydrogen is now a stone's throw from being competitive with oil. Only ten years ago, it cost $24/kg. Today we are somewhere between $4 and 5.5/kg. And we can get it down to $2/kg – the tipping point at which it becomes competitive with fossil fuels - within five years (at least in some regions) simply by kick-starting the first hydrogen projects. This is within reach. The technology exists. Large pilot projects are testing hydrogen in all kinds of applications. There is unprecedented commitment to full decarbonisation, and policymakers the world over are looking to give hydrogen the judicious nudges it needs.

We are on the cusp of a hydrogen revolution.

If I have convinced you that the time of hydrogen is near, then perhaps I have contributed to bringing it a little nearer. New markets are all about confidence. If producers can see that there will be a big market, they will invest in production.

If people can see that cheap green hydrogen will soon be available, they will invest in the technology to use it. There is a lot of pent up demand waiting in the wings. Trillions of dollars of consumer spending are waiting for green products to buy, and trillions of dollars of green funds are looking for projects to finance, but companies are holding off on major investment decisions because they are waiting to see which low-carbon technologies will come through. As we get a good line of sight on hydrogen's tipping point, there will be a rush to invest and hydrogen will snowball.

I don't mean to suggest that hydrogen is a slam dunk. It isn't – and neither is getting the whole world to net zero by 2050. The amount of renewable and low-carbon power we will need to install is dizzying. The investments we will need to mobilise are staggering. The choices and decisions we will all need to make are beyond number. But it is doable. And we should tackle these challenges with optimism, courage and determination because that's the only way we will get the job done.

I am writing this final chapter in Rome, where we still make daily use of ancient roads and aqueducts. Building them was a gargantuan effort at the time, but 2,000 years later we can safely say it was well worth it. We need a similar level of ambition today to build our own monumental infrastructure project: a forever energy system, inexhaustible and clean, united by hydrogen.

NOTES

INTRODUCTION

1 The Paris Agreement, a legally binding international treaty on climate change, was adopted by 196 Parties at COP 21 (the twenty-first meeting of the Conference of the Parties) of the United Nations Framework Convention on Climate Change (UNFCCC) in Paris on 12 December 2015, and entered into force on 4 November 2016. The United States ratified the agreement on 3 September 2019, signed the agreement by executive order under President Barack Obama on 22 April 2016, announced its intention to withdraw from it on 1 June 2017, during Donald Trump's tenure as president, and officially rejoined on 20 January 2021, under President Joe Biden.

2 Verne, J. (1927). *The Mysterious Island*. New York: Scribner.

3 'We Could Power The Entire World By Harnessing Solar Energy From 1% Of The Sahara'. (2016). *Forbes*. https://www.forbes.com/sites/quora/2016/09/22/we-could-power-the-entire-world-by-harnessing-solar-energy-from-1-of-the-sahara/?sh=1d62873d4406.

4 New Energy Outlook 2020. *BloombergNEF*. https://about.bnef.com/new-energy-outlook.

1 NOTHING, FAST

1 Diamond, J. (2005). *Collapse*. New York: Penguin.

2 '9 Out of 10 People Worldwide Breathe Polluted Air', World Health Organization. www.who.int/news-room/air-pollution. That's not including the nearly four million more estimated to die from cooking-fire

263

pollution. One 2021 study puts annual deaths from ambient pollution at a staggering 8.7 million. See Vohra, K., Vodonos, A., Schwartz, J., Marais, Eloise A., Sulprizio, Melissa P. and Mickley, Loretta J., 'Global Mortality from Outdoor Fine Particle Pollution Generated by Fossil Fuel Combustion: Results from GEOS-Chem'. *Science Direct* 195 (April 2021). www.sciencedirect.com/science/article/abs/pii/S0013935121000487.

3 Global Energy Review 2020. (2020). *IEA*. https://www.iea.org/reports/global-energy-review-2020.

4 Elaboration of World Resources Institute data.

5 'New Energy Outlook 2020'. (2020). *BloombergNEF.*

6 Analysis was only carried out for Italy, France, Spain. Forklift, ground transport and long-haul busses & trucks are added to every port and city cluster. Single refineries or fertilizer plants were not highlighted. Illustrative sizes are based on approximate average sizes of industrial facilities and transport hubs. Sources: SYSTEMIQ analysis for Energy Transitions Commission (2020) based on public sources retrieved November 2020: European Environment Agency, "European Pollutant Release and Transfer Register"; Fertiliser Europe, "Map of major fertilizer plants in Europe"; Eurofer, "Where is steel made in Europe?"; European Commission, "TENTec Interactive Map Viewer" and "Projects of common interest – Interactive map"; Gie, "Gas Infrastructure Europe"; CNMC, ""General Overview of Spanish LNG Sector"; McKinsey, "Refinery Reference Desk – European Refineries"; Fractracker Alliance, "Map of global oil refineries".

7 Damon Matthews, H., Tokarska, K. B., Rogelj, J. et al. (2021). 'An integrated approach to quantifying uncertainties in the remaining carbon budget'. *Commun Earth Environ*, 2:7. https://doi.org/10.1038/s43247-020-00064-9.

8 Global Energy Review: CO2 Emissions in 2020. (2021). *IEA*. https://www.iea.org/articles/global-energy-review-co2-emissions-in-2020.

9 Press Release: 'Global carbon dioxide emissions are set for their second-biggest increase in history'. (2021). *IEA*. https://www.iea.org/news/global-carbon-dioxide-emissions-are-set-for-their-second-biggest-increase-in-history.

10 Masson-Delmotte, V., P. Zhai, H.-O. Pörtner, D. Roberts, et al. (eds.). (2018). 'Summary for Policymakers'. *IPCC*. https://www.ipcc.ch/site/assets/uploads/sites/2/2019/05/SR15_SPM_version_report_LR.pdf.

11 *Ibid.*

12 We also need to address agriculture, land-use change, and waste management and treatment, but that is outside the scope of this book.

2 THE DAY EVERYTHING STOPPED

1 Global Energy Review: CO2 Emissions in 2020. (2021). *IEA.* https://www.iea.org/articles/global-energy-review-co2-emissions-in-2020

3 FEET OF CLAY

1 Seaborg, Glenn T. (1996). *A Scientist Speaks Out: A Personal Perspective on Science, Society and Change.* River Edge, NJ: World Scientific Publishing, 177.
2 Adopted 11 December 1997, came into force on 16 February 2005.
3 CO2 Emissions from Fuel Combustion: Overview. (2020). *IEA.* https://www.iea.org/reports/co2-emissions-from-fuel-combustion-overview.
4 Franzen, J. (2019). 'What if we stopped pretending?'. *The New Yorker.*
5 Szenasy, S. interview with McDonough, W. (2016). 'Why Architects Must Rethink Carbon (It's Not the Enemy We Face)'. *Metropolis.* https://www.metropolismag.com/cities/why-architects-must-rethink-carbon-its-not-the-enemy-we-face.
6 [1] BRENT average 2020, [2] TTF average 2020, [3] Coal ARA average 2020, [4] Average number, highly dependent on natural gas cost, [5] Assuming current renewable and electrolyser costs in various locations, [6] Average number, highly dependent on the cost of natural gas and assumed investment for CCS.
7 Marshall, G. (2014). *Don't Even Think About It.* Bloomsbury.
8 Franzen, J. (2019). 'What if we stopped pretending?'. *The New Yorker.*
9 Reynolds, P. (2001). 'Kyoto: Why did the US pull out?'. *BBC.* http://news.bbc.co.uk/1/hi/world/americas/1248757.stm.

4 THE STARS ARE ALIGNING

1 Ford, J. (2019). 'Net zero emissions target requires a wartime level of mobilisation'. *Financial Times.* The European Hydrogen Backbone. (2020). https://www.ft.com/content/412eea06-8eb7-11e9-a1c1-51bf8f989972.
2 Harari, Y. N. (2011). *Sapiens.* Penguin Random House.
3 Larry Fink's 2021 letter to CEOs. https://www.blackrock.com/corporate/investor-relations/larry-fink-ceo-letter.
4 Roser, M. (2020). 'Why did renewables become so cheap so fast? And what can we do to use this global opportunity for green growth?'. *Our World in Data.* https://ourworldindata.org/cheap-renewables-growth.

5 2019 figure from European Commission. 'Report from the Commission to the European Parliament, the Council, the European Economic and Social Committee and the Committee of the Regions: Energy Prices and Costs in Europe', EUR-Lex, Doc. No. 52019DC0001. (2019). https://eur-lex.europa.eu/legal-content/EN/TXT/?uri=CELEX:52019DC0001.

6 Hart, D. (2020). 'The Impact of China's Production Surge on Innovation in the Global Solar Photovoltaics Industry'. *ITIF*. https://itif.org/publications/2020/10/05/impact-chinas-production-surge-innovation-global-solar-photovoltaics.

7 Roser, M. (2020). 'Why did renewables become so cheap so fast? And what can we do to use this global opportunity for green growth?'. *Our World in Data*. https://ourworldindata.org/cheap-renewables-growth.

8 Accenture Report 'Lighting the Path: the next stage in utility scale solar development'https://www.accenture.com/_acnmedia/PDF-97/Accenture-Utility-Solar-Scale-POV.pdf.

9 Ford, N. (2021). 'Spain's record wind prices fail to curb the rise of solar'. *Reuters*. https://www.reutersevents.com/renewables/wind/spains-record-wind-prices-fail-curb-rise-solar.

5 IMBALANCE OF POWER

1 Schonek, J. (2013). 'How big are power line losses?'. *Schneider Electric Blog*. https://blog.se.com/energy-management-energy-efficiency/2013/03/25/how-big-are-power-line-losses/#:~:text=The%20transmission%20over%20long%20distances,as%20heat%20in%20the%20conductors.&text=The%20overall%20losses%20between%20the,range%20between%208%20and%2015%25.

2 Edwards-Evans, H. (2021). 'UK power system balancing costs down 15% on month in Dec 2020'. S&P Global. https://www.spglobal.com/platts/en/market-insights/latest-news/natural-gas/011821-uk-power-system-balancing-costs-down-15-on-month-in-dec-2020.

3 Policy Research Working Paper 8899. (2019). 'Underutilised potential: the business costs of unreliable infrastructure in developing countries.' http://documents1.worldbank.org/curated/en/336371560797230631/pdf/Underutilized-Potential-The-Business-Costs-of-Unreliable-Infrastructure-in-Developing-Countries.pdf.

4 Meyer, G. (2021). 'Energy grids target upgrades for zero carbon transition'. *Financial Times*. https://www.ft.com/content/eb7d651b-7d0a-4bb8-9a6d-8f5088b36c9b.

5 Entso-E Statistical Factsheet 2018, Entso-G Demand Data 2018.
6 Hearne, R. (2013). 'Gas Articles: What size Gas Meter do I need?'. 1Gas. https://1gasconnections.co.uk/what-size-gas-meter-do-i-need.

6 ELEMENTAL ENERGY

1 In the core of our sun, 600 million tons of hydrogen nuclei are fused into helium every second, releasing prodigious energy that slowly makes its way from the core to the sun's surface, heating it to a temperature of 5,800 K. The Earth, 92 million miles away, basks in its warmth. Temperature and density drive the fusion processes keeping the sun alight. High temperature is important because positively charged protons repel one another, and they can only collide with enough force to fuse together if they are moving very fast (that is, if the plasma is very hot). The high density at the sun's core, meanwhile, means that protons are colliding very often, so the reaction proceeds rapidly. This hardly sounds like the sort of engine we could ever hope to build ourselves, and yet nuclear fusion has been studied in labs on Earth since the 1930s, and experimental fusion reactors were up and running by the 1950s. Fusion reactors like ITER in Cadarache, southern France, can reach temperatures even hotter than the centre of the sun – up to a couple of hundred million degrees. At these temperatures, they can't achieve the high densities that prevail under the crushing pressure of the sun's core, but they try to keep the plasma contained as tightly as possible using magnets: a kind of high-stakes, high-status version of that game where you try to squeeze a balloon between your hands. If we can ever crack the problem of controlled nuclear fusion, we'll have limitless, low-pollution energy in vast quantities. But that vision is still a long way off.

2 Sternberg, S. P. K. and Botte, G. G. 'Fuel Cells in the Chemical Engineering Curriculum.' Department of Chemical Engineering, University of Minnesota Duluth. http://www.asee.org/documents/sections/north-midwest/2002/Sternberg.pdf.

7 FLIGHT OF FANCY

1 Monck Mason. *Aeronautica, or sketches illustrative of a theory and practice of Aerostation.* Westley, 1838.

8 UNFAIR COMPETITION

1 There are some pockets of natural hydrogen, and a few of these are being exploited. See, for example, Enu Afolayan. 'Hydrogen Power in Mali', Africa–Middle East: News and Perspectives from a Land of Opportunities. https://africa-me.com/hydrogen-power-in-mali.

2 Dukes, J. S. (2003). 'Burning Buried Sunshine: Human Consumption of Ancient Solar Energy'. *Climatic Change* 61:31–44. https://core.ac.uk/download/pdf/5212176.pdf.

3 Energy Brainpool elaboration World Energy Outlook 2019. *IEA*.

4 Global CCS Institute. (2017). 'Global Costs of Carbon Capture and Storage'.

5 Keith, D. W., Holmes, G., St. Angelo, D. and Heidel, K. (2018). 'A Process for Capturing CO_2 from the Atmosphere'. *Joule*.

6 Fasihi, M., Efimova, O., and Breyer, C. (2019). Techno-economic assessment of CO_2 direct air capture plants'. *J. Clean. Prod.*, 224:957–80.

7 Bastien-Olvera, B. and Moore, F. C. (2020). 'Use and non-use value of nature and the social cost of carbon'. *Nature Sustainability*, 4:101–8.

9 CLEANER COLOURS

1 Office of Nuclear Energy. (2020). 'Could Hydrogen Help Save Nuclear?'. Department of Energy. https://www.energy.gov/ne/articles/could-hydrogen-help-save-nuclear.

2 Adam Baylin-Stern and Niels Berghout. (2021). 'Is Carbon Capture Too Expensive?,' International Energy Agency. See: www.iea.org/commentaries/is-carbon-capture-too-expensive.

3 CCUS in Clean Energy Transitions: A new era for CCUS (2020). *IEA*. https://www.iea.org/reports/ccus-in-clean-energy-transitions/a-new-era-for-ccus.

4 Geißler, T., Abánades, A., Heinzel, A. et al. (2016). 'Hydrogen production via methane pyrolysis in a liquid metal bubble column reactor with a packed bed'. *Chemical Engineering Journal*, 299:192–200. https://doi.org/10.1016/j.cej.2016.04.066.

5 See 'Technology'. SG H2 Energy. https://sg-h2.squarespace.com/technology.

10 HANDLING HYDROGEN

1 'RH2-The Ultimate Decarbonizer'. RH2C. www.renewableh2canada.ca/rh2.html.
2 The European Hydrogen Backbone. (2020).
3 Andersson, J. and Grönkvist, S. (2019). 'Large-scale storage of hydrogen'. *International Journal of Hydrogen Energy*, 44:23:11901–19. https://doi.org/10.1016/j.ijhydene.2019.03.063.

12 THE POWER COUPLE

1 Yet, already intermittent renewables are leading to some pretty wild fluctuations in energy cost. In the UK, on 6 January 2021, high demand and low supply meant evening power prices were £1,500/MWh. A year earlier, in February 2020, the system was oversupplied thanks to strong winds from Storm Dennis, and the price crashed into negative territory at -£60/MWh. In 2020, there were fifteen separate occasions when UK power producers had to pay someone to take their power.

13 THE NEW OIL

1 Peel, M. and Fleming, S. (2021). 'West and allies relaunch push for own version of China's Belt and Road. *Financial Times*. https://www.ft.com/content/2c1bce54-aa76-455b-9b1e-c48ad519bf27.
2 Criddle, C. (2021). 'Bitcoin consumes more energy than Argentina'. *BBC News*.
3 At less than $1/MWh some Middle Eastern oil production and some Russian gas production will always be cheaper than renewables anywhere in the world.
4 Stein, E. V. et al. (2020). 'Fertility, mortality, migration, and population scenarios for 195 countries and territories from 2017 to 2100: a forecasting analysis for the Global Burden of Disease Study'. *Lancet*. https://www.thelancet.com/journals/lancet/article/PIIS0140-6736(20)30677-2/fulltext.
5 Radowitz, B. (2020). 'World's largest hydro dam "could send cheap green hydrogen from Congo to Germany"'. *Recharge News*. https://www.rechargenews.com/transition/worlds-largest-hydro-dam-could-send-cheap-green-hydrogen-from-congo-to-germany/2-1-871059.

6 www.statista.com.
7 'Will Australia's "hydrogen road" to Japan cut emissions?' *The Finance Info*, 2020, https://thefinanceinfo.com/2020/11/29/will-australias-hydrogen-road-to-japan-cut-emissions/

14 OUR GREEN MATERIALS

1 Gates, B. (2019). 'Here's a Question You Should Ask About Every Climate Change Plan'. GatesNotes. www.gatesnotes.com/Energy/A-question-to-ask-about-every-climate-plan.
2 'Iron and Steel', International Energy Authority tracking report, June 2020, https://www.iea.org/reports/iron-and-steel.
3 'Global Consumption of Plastic Materials by Region 1980-2015', Plastic Insight.
4 (2019). 'Mission Possible sectoral focus: plastics'. Energy Transitions Commission. https://www.energy-transitions.org/publications/mission-possible-sectoral-focus-plastics/
5 Cormier, Z. 'Turning carbon emissions into plastic'. *BBC Earth*. https://www.bbcearth.com/blog/?article=turning-carbon-emissions-into-plastic.
6 Roberts, D. (2020). 'The hottest new thing in sustainable building is, uh, wood'. *Vox*.
7 United Nations Food and Agriculture Organization. (2009). '2050: A Third More Mouths to Feed'. see: www.fao.org/news/story/en/item/35571/icode

15 A WARM GLOW

1 Cho, R. (2019). 'Heating Buildings Leaves a Huge Carbon Footprint, But There's a Fix For It'. Columbia Climate School. https://news.climate.columbia.edu/2019/01/15/heat-pumps-home-heating.
2 In Italy in a given year 0.85% of total building stock is renovated, Strategia per la riqualificazione enegetica del parco immobiliare nazionale (STREPIN), Ministero dello Sviluppo Economico, Novembre 2020.
3 Future of gas event, 21 January 2021.
4 The earliest biofuel experiments were as likely to produce hydrogen as methane. D. D. Jackson and J. W. Ellms demonstrated hydrogen production by microalgae (Anabaena) as early as 1896. Enzymes that split hydrogen gas are common, having evolved at least three separate times in the history of life on earth. Today's anaerobic digesters can also produce

hydrogen, if we want them to. In a fermenter producing methane, the hydrogen produced by one group of organisms is converted to methane by a second group. To make hydrogen, you just need to selectively inhibit the activity of second group of organisms by changing the pH and temperature.

5 In the UK, Worcester Bosch already have these boilers in production. https://www.boilerguide.co.uk/articles/switching-hydrogen-gas-grid-viable-option.

6 H21 Leeds City Gate. https://www.h21.green/projects/h21-leeds-city-gate.

7 Day, A. (2017). 'Sustainable Futures: Lighter than Air'. https://anthony-day.blogspot.com/2017/11/lighter-than-air.html.

16 GOING GREEN

1 FCHEA. (2019). 'Comments on Transportation and Climate Initiative Framework for a Draft Regional Proposal'. https://www.transportation-andclimate.org/sites/default/files/webform/tci_2019_input_form/TCI%20MOU%20Response%20FCHEA%202020-2-28.pdf.

2 'Bush Touts Benefits of Hydrogen Fuel', CNN, 6 February 2003, https://edition.cnn.com/2003/ALLPOLITICS/02/06/bush-energy

3 Even as this book went into production, batteries capable of fully charging in five minutes were produced in a factory for the first time by the Israeli company StoreDot. These batteries will only carry a car 100 miles or so, and require much higher-powered chargers than used today – but these are the kinds of problems that time and effort may fix – even by the time you read this book. The battery replaces the graphite in its electrodes with semiconductor nanoparticles based on the rare-earth element germanium, which isn't great for the environment. But StoreDot's plan is to replace the germanium with silicon, which is much cheaper. If they can do that (and prototypes are expected in 2021) then their fast-charge batteries really could be a game changer, and the electric vehicle landscape will be reset again.

4 It is estimated that there could be 50 million to 70 million electric vehicles (EVs) and plug-in electric vehicles (PHEVs) on the road in Europe by 2030, and that this will requires an investment of €375–475 billion in distribution grids. On the positive side, this will create jobs. See 'Making Power Grids Fit for the Transition Will Create 500,000 Jobs'. (2021). Eurelectric. www.eurelectric.org/connecting-the-dots.

5 Agence France-Presse. (2018). 'Germany launches world's first hydrogen-powered train'. *Guardian*. www.theguardian.com/environment/2018/sep/17/germany-launches-worlds-first-hydrogen-powered-train.

6 Keating, C. (2020). '"This is not a bus plan": Wrightbus' Jo Bamford's vision for catalysing the UK's hydrogen ecomomy'. *Business Green*.

7 Buckland, K. (2019). 'Explainer: Why Asia's biggest economies are backing hydrogen fuel cell cars'. *Reuters*. https://www.reuters.com/article/us-autos-hydrogen-explainer-idUSKBN1W936K.

8 Wayland, M. (2021). 'General Motors partners with Navistar to supply fuel-cell technology for new semitruck'. *CNBC*. https://www.cnbc.com /2021/01/27/general-motors-partners-with-navistar-to-supply-fuel-cell-technology-for-new-semitruck.html.

18 REACHING FOR THE SKY

1 British Airways Press Release. (2019). 'British Airways One Step Closer to Powering Future Flights by Turning Waste into Jet Fuel'. https://mediacentre.britishairways.com/pressrelease/details/   86/2019-319/11461.

2 CORDIS: Liquid Hydrogen Fuelled Aircraft – System Analysis (CRYOPLANE). https://cordis.europa.eu/project/id/G4RD-CT-2000-00192/it.

20 THE SAFE SIDE

1 Altmann, M. and Graesel, C. (1998). 'The acceptance of hydrogen technologies'. https://www.osti.gov/etdeweb/biblio/20584244

2 Markandya, A. and Wilkinson, P. (2007). 'Electricity generation and health', *Lancet*, 370:979–990.

3 Kolodziejczyk, B. and Ong, W-L. (2019). 'Hydrogen power is safe and here to stay'. *World Economic Forum*.

4 *Ibid*.

21 THE MISSION

1 Hydrogen Council, McKinsey & Company (January 2021): Hydrogen Insights 2021.

2 IRENA. (2021). 'Renewable Capacity Highlights'.

3 'New Energy Outlook 2020'. *BloombergNEF*.

4 According to BNEF – Hydrogen Economic Outlook – March 2020, 145 $/tCo2 will be required to bring hydrogen-based fuels at par with fuel oil in shipping even when hydrogen will cost $1/kg.

5 ITM Power, maker of PEM electrolysers and De Nora Industrie, which make the coating for alkaline electrolysers.

6 'Green Hydrogen: Time to scale up.' (2020). Bloomberg NEF. https://www.fch.europa.eu/sites/default/files/FCH%20Docs/M.%20Tengler_ppt%20%28ID%2010183472%29.pdf

7 IRENA. (2020). 'Green Hydrogen Cost Reduction: Scaling up Electrolysers to Meet the 1.5°C Climate Goal', International Renewable Energy Agency, Abu Dhabi

8 'Green Hydrogen: Time to scale up.' (2020). BloombergNEF. https://www.fch.europa.eu/sites/default/files/FCH%20Docs/M.%20Tengler_ppt%20%28ID%2010183472%29.pdf

9 'Breakthrough Strategies for Climate-Neutral Industry in Europe', Agora Energiewende, Wuppertal Institute, 2020. https://www.agora-energiewende.de/en/publications/breakthrough-strategies-for-climate-neutral-industry-in-europe-summary/.

22 CATAPULT COMPANIES

1 Yvkoff, L. (2019). 'In Battery Vs. Hydrogen Debate Anheuser-Busch Shows There's Room for Both Technologies with Nikola-BYD Beer Run'. *Forbes.* https://www.forbes.com/sites/lianeyvkoff/2019/11/22/anheuser-busch-demonstrates-theres-room-for-both-technologies-in-battery-vs-hydrogen-debate.

23 CALLING THE COPs

1 The views stated herein do not necessarily reflect the views of Goldman Sachs.

2 CCFDs pay out the difference between the price of emissions allowances (EUAs) and the contract price, thus effectively ensuring a guaranteed carbon price for the project. In exchange for this insurance, investors are liable for payment if the carbon price exceeds the contract's strike price.

3 Rajnish Singh, 'Creating Green Energy Partners with North Africa,' The Parliament, November 6, 2020, www.theparliamentmagazine.eu/news/article/green-energy-partners.

4 Liebreich, M. (2020). 'Separating Hype from Hydrogen – Part One: The Supply Side'. *BloombergNEF.* https://about.bnef.com/blog/liebreich-separating-hype-from-hydrogen-part-one-the-supply-side.

5 Green Car Congress. (2020). 'New Road Map to a US Hydrogen Economy'. https://www.greencarcongress.com/2020/03/20200322-h2map.html.

24 THE CONSUMER CAVALRY

1 Le Page, M. (2017). 'A load of bread emits half a kilo of CO2, mainly from fertiliser'. *New Scientist.*
2 (2018). 'Mission Possible: reaching net zero carbon emissions from harder -to-abate sectors by mid-century'. Energy Transitions Commission.

25 MAKING HYDROGEN HAPPEN

1 Goldman Sachs estimates the value of the global hydrogen market by 2050 at $11.7 trillion. See Goldstein, S. (2020). 'Green Hydrogen' Could Become a $12 Trillion Market. Here's How to Play It'. *Barron's.* www. barrons.com/articles/goldman-sachs-says-so-called-green-hydrogen-will-become-a-12-trillion-market-heres-how-to-play-it-51600860476.

ACKNOWLEDGEMENTS

I owe a debt of gratitude to Camilla Palladino, who has worked with me on this project from conception to publication. It was Camilla and her team who first showed me the model highlighting the importance of hydrogen to reaching full decarbonization (on that fateful afternoon in Milan in 2018), kicking off a conversation which has shaped Snam's strategy and contributed to the European policy discourse. I would also like to acknowledge Xavier Rousseau, Vieri Maestrini, Tatiana Ulkina, Fabrizio De Nigris and the whole strategy team at Snam for their contribution to this work.

I would like to thank Massimo Derchi, Cosma Panzacchi, Paolo Tosti and the technical division, Dina Lanzi, Marco Chiesa, Pere Margalef and many others in Snam for providing a wealth of information on hydrogen technologies and the challenges and opportunities which they may provide, which has enriched this book.

Early readers of the manuscript helped to weed out mistakes and suggest additions. I am grateful to Ermenegilda Boccabella, Jorgo Chatzimarkakis, David Hart, Thomas Koch-Blank, Markus Wilthaner and my colleagues Alessandra Pasini, Claudio Farina, Patrizia Rutigliano, Gaetano Mazzitelli, Salvatore Ricco and Laura Parisotto for their time, expertise, fresh eyes and sharp pencils. My mother, father and brother were also early readers, and I thank them for their encouragement (and lots more besides).

I am indebted to the many peers and industry executives who generously agreed to share their specific sectoral knowledge and experience, and to the many brilliant people with whom I have had the opportunity to discuss ideas on the energy transition and hydrogen. Among them Gabrielle Walker (who was also kind enough to read an earlier draft of this book and provide detailed feedback) and Catapult 'co-originators' Nigel Topping, Jules Kortenhorst and Lord Adair Turner. I am indebted in particular to Jonathan Stern and Chris Goodall who lent their time and expertise at the final hour.

You wouldn't be holding this book at all if it weren't for my agent, Peter Tallack, who took it on when it was little more than a collection of ideas written on scraps of paper, my editor, Izzy Everington, who is as enthusiastic about hydrogen as I am, TJ Kelleher and Eric Henney, who had lots of great suggestions, and the publishing teams at Hodder Studio and Basic Books. I would also like to gratefully acknowledge the support received in the crafting of this work from Stephen Battersby, Simon Ings and Tom Burke.

The Rocky Mountain Institute, Energy Transitions Commissions and Goldman Sachs were kind enough to give me permission to reproduce some great charts.

Finally, thank you to Selvaggia for bearing with me as I got engrossed in this project over weekends and some very late nights, and to Lipsi and Greta for being the reason why I care so much about what happens next.

BIBLIOGRAPHY

An Ocean of Air: A Natural History of the Atmosphere, Gabrielle Walker (Mariner Books, 2008)

Burn Out: The Endgame for Fossil Fuels, Dieter Helm (Yale University Press, 2017)

Climate of Hope: How Cities, Businesses, and Citizens Can Save the Planet, Michael Bloomberg and Carl Pope (St Martin's Press, 2017)

Con tutta l'energia possibile, Leonardo Maugeri (Sperling & Kupfer, 2011)

Designing Climate Solutions: A Policy Guide for Low-Carbon Energy, Hal Harvey, Robbie Orvis and Jeffrey Rissman (Island Press, 2018).

Don't Even Think About It: Why Our Brains Are Wired to Ignore Climate Change, George Marshall (Bloomsbury, 2015)

Drawdown: The Most Comprehensive Plan Ever Proposed to Reverse Global Warming, Paul Hawken; (Penguin, 2017)

Energy and Civilisation, Vaclav Smil (MIT Press, 2017)

European Hydrogen Backbone, Guidehouse (2021)

Gas for Climate: Gas Decarbonisation Pathways, Navigant (2020)

Gas for Climate: The Optimal Role for Gas in a Net-Zero Emissions Energy System, Navigant (2019)

Global Energy Transformation: A Roadmap to 2050,
International Renewable Energy Agency (2019)

How to Avoid a Climate Disaster, Bill Gates (Allen Lane 2021)

Hydrogen Decarbonisation Pathways, The Hydrogen Council (2021)

Hydrogen Economic Outlook, Bloomberg New Energy Finance (2020)

Hydrogen Insights 2021: A Perspective on Hydrogen Investment, Deployment and Cost Competitiveness, The Hydrogen Council (2021)

Hydrogen is the New Oil, Thierry Lepercq (Le Cherche Midi, 2019)

Hydrogen: The Economics of Production from Renewables, Bloomberg New Energy Finance (2019)

Losing Earth, Nathaniel Rich (Picador, 2019)

Making the Hydrogen Economy Possible: Accelerating Clean Hydrogen in an Electrified Economy, Energy Transitions Commission (2021)

Mission Possible: Reaching Net-Zero Carbon Emissions from Harder-to-Abate Sectors by Mid-Century, Energy Transitions Commission (2018)

Net Zero by 2050: A Roadmap for the Global Energy Sector, International Energy Agency (2021)

On Fire: The Burning Case for a Green New Deal, Naomi Klein (Simon & Schuster, 2019)

Six Degrees: Our Future On a Hotter Planet, Mark Lynas (National Geographic, 2008)

Sustainable Energy – Without the Hot Air, David JC MacKay (UIT Cambridge, 2009)

The Citizen's guide to Climate Success, Mark Jaccard (Cambridge University Press, 2020)

The Future of Hydrogen: Seizing Today's Opportunities,
International Energy Agency (2019)
The Future We Choose, Christiana Figueres (Knopf, 2020)
The Global Gas Report, International Gas Union, Snam,
Bloomberg New Energy Finance (2020)

The Hot Topic: What We Can Do About Global Warming,
Gabrielle Walker and David King (Bloomsbury, 2008)
The Hydrogen Economy, Jeremy Rifkin (Tarcher/Putnam,
2002)
The New Map: Energy, Climate, and the Clash of Nations,
Daniel Yergin (Penguin 2020)
The Prize: The Epic Quest for Oil, Money and Power, Daniel
Yergin (Simon & Schuster, 2008)
*The Tipping Point: How Little Things Can Make a Big
Difference,* Malcolm Gladwell (Black Bay Books, 2013)
The Uninhabitable Earth, David Wallace Wells (Random
House, 2019)
There is no Planet B, Mike Berners-Lee (Cambridge
University Press, 2019)
*Tomorrow's Energy: Hydrogen, Fuel Cells, and the Prospects for a
Cleaner Planet,* Peter Hoffmann (MIT Press, 2012)
What We Need To Do Now, Chris Goodall (Profile, 2020)

GLOSSARY

AC
Alternating current – electrical current that rapidly switches direction, typically 50 times per second. This makes it possible to shift voltage up and down using simple devices known as transformers.

Battery electric vehicle (BEV)
A vehicle powered by electric motors, using electrical energy stored in onboard batteries.

Blue hydrogen
Hydrogen made from natural gas, with the carbon dioxide byproduct then captured and stored.

Carbon dioxide (CO_2)
A gas consisting of one carbon and two oxygen atoms, which is generated by burning fossil fuels and making cement, as well as various natural processes. It is a greenhouse gas, and the largest contributor to human-induced climate change.

Carbon capture and storage (CCS)

Capturing CO2 and putting it into a store such as a disused oilfield. A related term is carbon capture, utilisation and storage (CCUS), which includes schemes to make use of the captured gas.

Carbon tax

A tax on goods and services based on the amount of greenhouse gases they release into the atmosphere.

Climate change

A broader term than global warming, climate change refers to shifts in any aspect of climate, including winds and rainfall patterns.

CO_2 equivalent

A way to compare and add up the effects of different greenhouse gas emissions. For a kilogram of any given greenhouse gas, its CO2 equivalent is the amount of CO2 that would produce the same amount of global warming. (The relative effect of a given gas is called its global warming potential. Because different gases linger in the atmosphere for different lengths of time, global warming potential has to be averaged over some timeframe, usually 100 years. Methane, for example, has a residence time of about 12 years, much shorter than CO2, but its immediate heating effect is so strong that methane's 100-year global warming potential is more than 30 times that of CO2.)

Dispatchable power

Electricity generation that can be rapidly turned on or off, to cover fluctuations in demand and fluctuations in the supply from intermittent power sources.

Distribution

In the energy industry, distribution is distinguished from transmission. Large cables at high voltage perform long-distance transmission of electricity. More localised networks of smaller wires at lower voltage then distribute the power to users. Likewise, the gas industry distinguishes large-diameter long-distance transmission pipelines from local distribution networks of smaller pipes.

Electrical current

The rate of flow of electrons (or other charge carriers). Its SI unit is the ampere or amp (A).

Electrical grid

The network connecting electricity producers to consumers. The physical grid includes electricity generators, transmission and distribution wires, substations where voltages are transformed, and end users such as homes and businesses. As well as the physical hardware, the term can also refer to the companies and structures that respond to supply and demand, including markets and regulation.

Electrolyser

A device that splits water into hydrogen and oxygen using electrical power.

Energy

A measure that quantifies amounts of work (such as lifting a weight against gravity) and heat. SI unit is the joule (J). Different forms of energy can be converted to one another. If you set light to some hydrogen, its chemical energy is converted into both thermal energy (heat) and mechanical energy (sudden motion, which you hear as a pop).

Fossil fuels
Substances formed from organic matter trapped underground and transformed by heat and pressure over millions of years. Oil, coal and natural gas are the most widely used fossil fuels. Burning them adds the greenhouse gas carbon dioxide to the atmosphere, contributing to climate change.

Fuel cell
A device that converts chemical energy into electricity. Hydrogen fuel cells react hydrogen with oxygen to generate electrical power, with water as the waste product.

Fuel cell electric vehicle (FCEV)
A vehicle powered by electric motors, with energy stored onboard in the form of hydrogen, which is converted to electricity using a fuel cell.

Gigawatt (GW)
One billion watts, a watt being the SI unit of power. A large power station may have a generating capacity of a few GW.

Gigatonne
One billion metric tonnes, a unit often used to measure carbon dioxide emissions.

Global warming
Earth's rising temperature. Natural temperature changes have occurred in the past, but the term is generally used to refer to

recent rapid warming caused by human-generated greenhouse
gas emissions.

Greenhouse gas
A gas that allows sunlight to enter Earth's atmosphere and then
prevents heat from escaping – acting like a greenhouse window.
The main greenhouse gases are water vapour, carbon dioxide
(CO_2), methane (CH_4), nitrous oxide (N_2O), ozone (O_3),
and various chlorofluorocarbons and hydrofluorocarbons.

Green hydrogen
Hydrogen made using renewable electricity to split water in an
electroyser.

Hydrogen
The first element in the periodic table. A hydrogen atom is
formed of one proton and one electron. Hydrogen gas is
usually in the form of H_2 molecules, pairs of hydrogen atoms
bound together.

Intermittent energy source
A source with power output that depends on natural variabil-
ity. Solar and wind energy are both intermittent, as they depend
on the weather.

Kilowatt hour (kWh)
A unit of energy, equal to a power of 1000 watts sustained for
one hour. One kWh equals 3.6 megajoules.

Learning rate
A measure of how rapidly costs fall as a particular technology

scales up. Specifically, the percentage fall in cost produced by a doubling in demand.

LH2/LOX
Liquid hydrogen / liquid oxygen

Megawatt hour (MWh)
A unit of energy, equal to a power of one million watts sustained for one hour. One MWh equals a thousand kWh.

Mtoe
A unit of energy, standing for million tonnes of oil equivalent. One Mtoe equals 11.63 million megawatt hours (MWh).

Natural gas
A flammable mixture of gases that occurs naturally in underground deposits. It is mainly methane, with varying proportions of other gases including ethane and other more complex hydrocarbons, hydrogen sulfide, carbon dioxide and nitrogen. It is used as fuel for heating, cooking and electricity generation, and as a chemical raw material in industry.

Net zero
The goal of bringing greenhouse gas emissions down, and/or increasing the amount we actively remove from the atmosphere, so that emissions balance removals – i.e. net emissions are zero.

Particulate matter
Small particles of solid or liquid droplets, commonly emitted by burning fossil fuels. PM2.5 is a measure of the amount of

particulate matter with diameter below 2.5 micrometres, and it is the most destructive form of air pollution worldwide, causing millions of deaths per year.

Photovoltaic (PV) cell

A device that converts sunlight directly into electricity, through the photoelectric effect. Often called a solar cell.

Renewable energy

Energy sources that are naturally replenished, such as solar, wind, hydro, biomass and geothermal.

Smart grid

An electrical grid that uses information in a sophisticated way to balance supply and demand (for example with smart meters that send data to energy suppliers, and appliances that respond to changing circumstances). Smart grids can be more reliable, efficient and green than traditional electrical grids, although they bring some concerns over privacy and security.

Smog

Dense air pollution, a haze of particulate matter. A common form of smog in the modern world is photochemical smog, produced by chemical reactions driven by sunlight, mainly involving hydrocarbons and nitrogen oxides from vehicle exhaust.

Solar energy

Radiant energy from the Sun, which is mostly in the form of visible and near infrared light.

Sustainability
Keeping a balance between human consumption and the planet's resources, especially the regenerative power of nature, so that our activity does not threaten the ability of the planet to support us and other species. According to the UN World Commission on Environment and Development, 'sustainable development ... meets the needs of the present without compromising the ability of future generations to meet their own needs.'

Voltage
A measure of how strongly an electrical current is driven, analogous to pressure in a water-filled pipe. Also known as potential difference and electromotive force. SI unit is the volt (V).

Watt (W)
The SI unit of power, how rapidly energy is transferred or used. One watt of power is equal to a current of one ampere moving across a voltage of one volt.

Wind farm
A collection of wind turbines used to generate electricity.

APPENDIX

A note on the heating value of hydrogen

The energy contained in a fuel cannot all be used in real world applications. Losses come from the conversion process and exact comparisons require careful and specific analysis. Because of this, different fuels are usually compared at a high level on the basis of their 'heating value'. Even this has different incarnations. The convention used is that higher heating value (HHV) assumes that heat energy in any water vapour produced is captured as the vapour is condensed. In contrast the lower heating value (LHV) is used where this heat is not captured.

For hydrogen, if you use it in a fuel cell no water vapour is usually produced so the HHV is a good representation of its energy content. If the hydrogen is burnt in a turbine and steam is not recovered, the LHV would be the appropriate measure. Of course, where hydrogen is compared to another fuel (e.g. natural gas), this would need to be considered in the same way.

In this book, for ease of comparison, we use the HHV throughout. For hydrogen this is nearly 40kWh per kilogram. The LHV would be 33kWh per kg. Comparison with other fuels is on the same basis.

What can I do with 1 kg of hydrogen?	
Activity	*Number (#)*
Kilometres with a car	90
Kilometres with a truck	15
Washing machine runs	10
Dishes of pasta	30
Showers	20
Watching football matches on TV	180
Charging my smartphone	1,200

What do I need to produce 1 kg of Hydrogen?	
Activity	*Value*
Energy	56.3 kWh
Electrolysis and PV capacity	35 W
Land (for solar PV)	0.9 m^2
Water	9 l

How much hydrogen would we need todecarbonize each sector in 2050?	
Sector	Hydrogen (million tonnes/year)
Steel	122
Cement	47
Other Industry	43
Petrochemicals	32
Power	244
Heavy Trucks	92
Cars	42
Light trucks and buses	34
Shipping	23
Rail	12
Residential	69
Commercial	41
Total	**801**

What resources would we need to decarbonize all sectors in 2050?	
Activity	Value
Energy	45 TWh
Electrolysis and PV capacity	28.2 GW
Land (for Solar PV)	700 km^2
Water	7.2 billion m^3

Global macro-economic and energy data, 2019	
World gross domestic product (GDP)	$90 trillion
World Population	7.7 billion
Annual energy consumption (global)	170,000 TWh
Annual energy expenditure (global)	$7 trillion
Average energy cost	$40/MWh
Energy spend as a proportion of GDP	8.2%
Annual energy consumption per capita	22 MWh
Annual energy expenditure per capita	$900

Credit: Snam

Marco Alverà began his career working at Goldman Sachs in London before moving to Enel, the world's largest renewable-energy company, and subsequently Eni, the oil and gas major. Since 2016, he has served as CEO of Snam, Europe's largest gas pipeline company. He lives in Milan.